Contents

References in the text to suppliers listed in Appendix B are given by page number and individual reference number, eg (p181 3.1).

Introduction

This is a handbook of footpath management. It has been written for conservation volunteers and others who are involved in the design, construction, maintenance and repair of paths.

The term 'footpath' is used in its non-legal sense to include any path or track that is mainly or only used by people on foot, whether a public right of way or a private path. Although mainly concerned with rural paths on farmland, moorland, mountain, woodland and coast, this handbook is also relevant to the the management of informal paths in urban areas.

The assets

England and Wales have over 120,000 miles of public rights of way, and many other areas to which the public have access. Scotland too has many miles of public rights of way, although definitive maps showing them are not required by law (see Chapter 2). Large areas of mountain and moorland are generally accessible to the public.

The path network is one of Britain's greatest recreation resources. To quote a former Minister of Housing and Local Government (from Allan Patmore, 1970), "everybody's idea of the English countryside is the right to walk across a field and sit under a tree, to walk along a track through unspoilt woodlands or over fells, to sit by a lake and fish, or to take the dog for a walk across the common .... in Britain one can be part of the countryside."

The problems

With this great asset come many problems. Rights of way are the responsibility of landowners and local authorities, and between them much is left undone through lack of resources, time or will. For many landowners, the right of people to walk across their land is an interference to farming, forestry or game, as well as wildlife. Many landowners take good care of their paths, gates and stiles, knowing that walkers are much more likely to keep to them if they are waymarked and easy to follow. It is the neglected paths that cause trouble both to walker and landowner. Some local authorities have permanent staff employed full-time on maintaining paths, treating them correctly as the simplest class of highway. Others virtually ignore them.

Another great problem is the uneven spread of use over the path network. It is ironic that a tiny percentage of paths are becoming eroded and unsightly through over-use, while the remainder become difficult to follow through lack of use.

This is not surprising. Most people want to walk to the spectacular places, to hill and mountain tops, coastal cliffs, lakes and waterfalls. The attractions of certain areas have been increased by the designation of long distance paths, and by publicity. This is part of the complex subject of recreation management, which is beyond the scope of this publication. This handbook is concerned with the state of paths today, and what can be done by practical means to improve them. There are two main areas of work. Firstly, the opening up of neglected paths, and secondly, the repair of paths damaged by over-use. Both are equally important.

It may well be asked why volunteers need become involved at all, when local authorities and land-owners are responsible for paths. The reason is the same one which motivates volunteers to become involved in many areas of life: because they care enough about the situation to take positive action themselves, and because they gain enjoyment and satisfaction from this involvement.

The good signs

There are signs that these problems are being tackled. The Ramblers' Association and many local amenity groups are involved in the legal protection and practical maintenance of rights of way. In their 1982 prospectus, the Countryside Commission have stated that, whilst maintaining the necessary support to the popular and over-used places, they will be giving increased support to schemes improving recreation and access in urban fringe and neglected land.

The price of petrol has discouraged many people from travelling long distances to uplands and National Parks, and instead they are finding interest in their home areas. The damage done by repeated field trips, sponsored walks and other youth activities in certain areas is being realised, and efforts are being made to spread the use more widely.

This handbook cannot hope to give all solutions for all areas, with the enormous variety of landscape, soils and climate that affect paths. There is a broad division in the subject, between the mainly 'upland' work of drainage and erosion control, and the 'lowland' work of clearance, waymarking and stile construction. This handbook is designed to be used as a reference for ideas and solutions to particular problems. It is hoped though that readers will find ideas in all chapters that are of interest, even if not of direct relevance to the problems they normally encounter. Examples of work have been taken from areas throughout Britain.

1 The Pattern of Paths

The network of paths and tracks which we can walk today has developed over thousands of years, as successive generations have made their way about the land, for hunting, fishing and farming, trade, military venture and pilgrimage, and for pleasure and recreation.

The following section gives a brief historical summary of the development of paths and tracks.

Paths and Tracks

Pre-Roman Britain

It is likely that the first routes were established by herds of animals, moving between salt marshes, lowland grazing and upland moors in the years following the end of the last glaciation (Taylor, 1979). These migratory herds would probably have chosen the easiest routes, along ridges, through sparse vegetation, crossing rivers where fordable, and taking the lowest passes through the mountains. Grazing and trampling along the route would have spread over a wide zone where the land allowed, but would have been confined to narrow tracks through difficult country. The people who lived at that time were also constantly on the move, being hunters, fishermen and gatherers. They followed the herds for meat and hides, and had similar needs for water and shelter, and so shared the same routes. These routes thus became established as trackways, the chalk ridgeways being amongst the earliest.

About 6,000 B.C., fire was first used to clear forest for agriculture, and permanent settlements were established. Trade developed, as evidenced by chert tools from Portland found in many sites in the South West. After 4,000 B.C., the advances of the Neolithic revolution resulted in large areas being cleared, and recent archaeological evidence is increasingly showing that settlement was not just confined to the dry uplands, but extended over much of the country, and that the Neolithic peoples were far more organised and sophisticated than had been previously thought.

There is every likelihood that these people knew their way around their hinterland intimately, for they had to, in order to survive. Evidence of trackways is hard to establish, as their use in successive generations makes them difficult to date. However, recent excavations for peat in the Somerset levels have revealed causeways of brushwood and timber made in a similar way to that recommended today. The earliest tracks of piles of brushwood date from 3,000 B.C. Later causeways of split timber have been dated to around 2,000 B.C. It has been calculated that one mile of such trackway would have needed 20 miles of split logs, and already 14 miles of causeway have been found. No wonder our efforts at making paths over deep peat seem very much less successful than those of the Neolithic people. It is interesting that ICI's solution to wet peat on the Pennine Way at the Snake Pass, a product called 'Paraweb', has been abandoned in favour of the Neolithic technique of brushwood bundles (see p78).

After 1,400 B.C., the population began to increase rapidly, and in north Bedfordshire and south Northamptonshire for example, it has been shown that there was at least one settlement in every square mile of countryside. Heavy wheeled carts were in use by 800 B.C., and Mediterranean wine jars found in East Anglia indicate the extent of trade. It is likely that in the centuries immediately B.C., the broad pattern of roads, tracks and paths in use at the beginning of the Industrial Revolution was already in existence.

The Roman Occupation

Britain has only seen two major and organised attempts at roadbuilding, these being during the 1st and 20th Centuries A.D. Between 43 and 47 A.D. the Romans occupied an area south-east of a line from Exeter to Lincoln, and later spread north and west. During the Roman occupation, about 10,000 miles of road were built, mostly within the first hundred years. All these roads were carefully surveyed, engineered and constructed. It is an awe-inspiring achievement, not least of which is the accuracy with which the direction of each route was established. The Foss Way, which runs for 200 miles along the Exeter-Lincoln frontier line, only deviates 8 miles from the direct line. The network was not only efficient, but extensive. In south-east Cambridgeshire for example, no settlement was further than seven miles from a surfaced road, and communication by road did not surpass this standard until the 20th Century.

Apart from their straightness, the main characteristic of Roman roads is that they were built on an embankment. This was called an agger, and could be from 300mm to 2 metres high, and 3 to 20 metres wide. It was made of the strongest material conveniently available, which could be gravel, flint, rammed chalk or simply earth. Material was taken either from a ditch alongside the agger, or from borrow pits. Both methods are still recommended for paths (see Chapter 8). The agger raised the road to keep it well drained, and helped to even out irregularities in the ground. The agger was then surfaced to make the road, usually for only part of its width. The surface was made of a base layer of large stones topped with gravel or small stones. In the Weald, slag from iron workings was sometimes used. Fords were made by tipping blocks of stone into the river bed. In other places, timber or stone bridges were built.

After the last Roman troops left in 407, the system of trade, communication and administration disintegrated, and many of the major routes fell into disuse.

The Mediaeval Period

Little is known about the early part of the Mediaeval period called the Dark Ages, which lasted from the departure of the Romans until the Norman conquest. Some parts of the Roman roads were probably used for local traffic, but most communication would have been along the local tracks that linked settlements. Saxon settlements were small and dispersed, with few villages or towns. The Dark Ages in myth and legend evokes a picture of primitive people struggling to survive in a wooded and hostile land. The reality was probably quite different, as shown by the records of cultivated land in the Domesday Book of 1086, which are only slightly below the acreage in cultivation in 1914. The landscape of tiny farms and fields with numerous interconnecting tracks can still be seen, mainly unchanged, in parts of Devon and Anglesey.

In other areas, particularly in the Midlands, the system of communal strip fields with outlying commons developed. Here there would have been unenclosed paths running between the strips, some giving access to the common land beyond. In the uplands and moorlands of the north and west where most land was not cultivated, tracks were used for driving animals, for access to cut peat and bracken, for mining, and for trade. Many of these tracks still exist today, some with names that indicate their past use such as Drovers' Ways, Miners' Ways, and Saltways.

As far as we know, little attempt was made to maintain or improve roads during the Mediaeval period, and sea and river travel was used as much as possible. In some parts of the country it was compulsory for serfs to do six days labour a year on the roads, but this was probably little more than removing obstructions and filling the worst holes. Mud and muck were dug out and used as manure, serving only to lower the tracks further and make water lie on them. Movement during winter was very limited, with stock being moved only when the land was reasonably dry.

Enclosures

Enclosure of common land and open fields occurred throughout the latter part of the Mediaeval period, but reached its peak between 1750 and 1850, when over 4,000 Acts of Parliament were passed. Each parish was done separately, with commissioners allotting land according to 'traditional rights'.

The paths which gave access to the strips of the open field were lost, and many tracks linking settlements and villages were re-aligned. As neighbouring parishes were not necessarily enclosed at the same time, the enclosure surveyors laid the track in as straight a line as possible to the boundary where the old route crossed it. When the same process occurred later in the neighbouring parish, a corner resulted. These corners and dog-legs can still be seen, marking the parish boundaries. Sometimes the surveyors set aside land for quarrying road material. Not all roads and tracks were re-aligned, as often there were not the funds available to build new roads on top of the cost of surveying, valuing and fencing.

New local paths were set out or developed by custom across the enclosed fields, linking village with church, farm and outlying settlement.

Turnpikes

At the same time as the Enclosures, other legal processes were happening which affected the system of paths. These were the Turnpike Trusts, the first founded by Act of Parliament in 1663, and with many following between 1700 and 1850. The Acts permitted the setting up of a Trust to construct and maintain a section of road in return for a levy from travellers. These Turnpike roads were a great improvement for travellers by horse and carriage, and so traffic was concentrated along them and other routes were little used. Thus the Turnpike Trusts helped turn the previous network of unimproved tracks into the hierarchical system of major and minor roads, tracks and paths that we see today.

Drove Roads

Drove roads from Wales to England had been in use since the 13th Century, but the trade from Scotland did not develop until after the Act of Union in 1707, when around 30,000 cattle a year were brought south over the border. The drove roads kept to high and unenclosed land, where there was space for the huge herds to graze and be kept together at night. Some of the drove roads were carefully constructed and surfaced. and examples remain with zig-zags, revetments, and

7

even steps on steep ground. The roads stayed in use throughout the era of the Turnpikes, the drovers preferring to keep to open ground and avoid paying tolls. However, the droving trade declined rapidly with the growth of the railway system in the 19th Century.

Wade's Military roads

Following the uprising in Scotland in 1715, military roads were built throughout the Highlands, after the recommendations of General Wade. These took over 70 years to complete, but were mostly not as well surveyed or built as the Roman roads in northern England. The Roman roads in the uplands make slow and easy ascents, with zig-zags where necessary, whereas General Wade's roads built by the British army tend to stay in the valleys as long as possible, with consequently very steep inclines at the heads of valleys and mountain passes. Some have become modern roads, others remain as tracks.

Improvement

Although the canals, and later the railways, took most of the long distance trade, many roads and tracks had increased traffic, generated by the railways. A few efforts were made to improve the road system, but it was not until 1894 that the responsibility for local roads was given to the Rural District Councils and work slowly began to meet the needs of the new motor transport. The decisions made in the 1920s and 1930s, as to which roads would be properly surfaced and which left as tracks, had far-reaching consequences for the development of rural communities. As shown throughout the history of communication, once a route is improved, it generates its own traffic and building development. Contact declined between neighbouring villages not joined by a surfaced road, making the pattern of social and economic movement that exists today. The tracks left unsurfaced are the byways and bridlepaths which are so important now for recreation.

Meantime, throughout the 18th, 19th and early 20th Centuries, the local paths crossing fields and woods were kept in use wherever people walked daily to work, to school, to shop, church and pub. Many of course were lost by urban and industrial development, and to the more efficient use of arable land.

Against the mainly economic use of paths, the need was increasing from the early 19th Century for people from urban areas to find peace and relaxation in the countryside, and these desires came to be voiced by pressure groups where disputes arose between landowners and walkers.

Rights of way

In 1826, the Manchester Society for the Preservation of Ancient Footpaths was formed to fight a legal battle against a landowner who tried to close the paths on his land at Flixton, near Manchester. Their success was followed by the founding of many similar groups, including the Scottish Rights of Way and Recreation Society in 1845, whose actions slowly built up the body of case law on public rights of way. However, there was no written law preventing a landowner closing a path, and such closures were only opposed if groups took legal action. The pressure for a legal status for public paths, combined with pressure for access to mountains, moorlands and commons (see below) eventually led to the legislation of the National Parks and Access to the Countryside Act 1949 (England and Wales), passed in the mood of social reform which followed World War II. The Ramblers' Association, founded in 1935, was very active in this movement.

One of the requirements of this Act was for local authorities in England and Wales to survey the course and status of all known public paths in their area, and publish the information in the Definitive map (see p11). In Scotland the rule of common law remained at that time unchanged.

Lands with Public Access

It is estimated that there are 1.2 million hectares of rural land in effective recreational use in England and Wales, comprising 8% of the total land area (Allan Patmore, 1970). This includes a large amount of land to which the public have access though not by right of way. On many of these areas the public can wander at will. The areas include commons, heaths, moors, mountains and coasts. Much of this is referred to as 'open country', because for reasons of soil, aspect or altitude it was never enclosed or cultivated. For the same reasons, there were few paths, and the only tracks that crossed these areas were the long distance drove roads and pack-horse trails, together with miners' tracks and other paths with special purposes.

Disputes over access first arose when the increasing demands for recreation from city-dwellers coincided with the fashion for grouse-shooting, and landowners tried to exclude the public to prevent disturbance to grouse. Similar disputes occurred over access to commons near urban areas, increasingly valued by townspeople for recreation.

MOUNTAINS AND MOORLANDS

Conflicts were strongest in the Peak District, where working people from cities such as Sheffield and Manchester sought fresh air and recreation on the open moorlands. Although action was taken in many cases to secure rights of way, the basic freedom sought was the 'freedom to roam'. This movement began with the formation of the Hayfield and Kinder Scout Ancient Footpaths Association in 1876. From 1888 and on many subsequent occasions private member's bills were introduced into the House of Commons to try and secure public access to all uncultivated mountain and moorland. Such legislation has still not been passed, although the National Parks and Access to the Countryside Act 1949 and the Countryside (Scotland) Act 1967 provide for the formation of access agreements.

Official access land

Under the National Parks and Access to the Countryside Act 1949, planning authorities in England and Wales were required to make an access survey of all open country in their area, which comprised mountain, moor, heath, down, cliff or foreshore. The authorities could then form agreements with landowners to secure public access, or make access orders if agreement could not be reached. In practice these powers have only seen limited use, of which most has been in the Peak District, where 19,328 hectares were subject to agreement by 1970.

In the Countryside Act 1968, the definition of 'open country' was extended to include woodlands and riverbanks, though few such agreements have yet been made.

Similar provision exists for access agreements in Scotland, under the Countryside (Scotland) Act 1967.

Unofficial access land

Much open country in England, Wales and Scotland is subject to 'de facto' access. 'De facto' is a term to describe a situation that exists, though not by legal right. 'De facto' access occurs by tradition, and often because it is not physically feasible for the landowner to prevent access. However, if use by the public increases or changes in pattern, it may become worthwhile for the owner to prevent access.

Unofficial access land is especially important in Scotland, where there are no definitive maps of rights of way, few commons, and only a few areas subject to access agreements.

COMMONS

It is estimated that there are 650,000 hectares of common land in England and Wales, of which a third is in Wales, and much of the remainder is in northern England. The type of land varies, but includes heath, grassland, scrub, woodland, moor and mountain.

Common land is the remnant of the manorial system of the Middle Ages. It is land owned by an individual, a company or a local authority, to which other people, called the commoners, have certain rights. These rights include, amongst others, the right to graze animals, to cut bracken, to fish or to collect firewood. Pressure for access by people other than the commoners was first voiced in the south of England, where landowners attempted to prevent public access to some of the commons around London. This led to the formation of the Commons Preservation Society in 1865, from which grew the National Trust, and the Commons, Open Spaces and Footpaths Preservation Society.

A major success was the Law of Property Act 1925 which gave the public right of access 'for air and exercise' to all commons in urban areas in England and Wales. Other commons only have legal public access if special agreements or Acts have been made, and these, together with the urban commons, only constitute about a quarter of the total area of common land. On other commons not subject to legal public access, it is often not worth the owners while, or he is not allowed, to fence it off and protect it from access by people other than the commoners. Thus, most commons are subject to 'de facto' access.

Some commons are now only used for recreation, and commoners rights are seldom exercised. Other commons, particularly in the uplands, are vital for the livelihood of those commoners with grazing rights. Whether legalised public access should be allowed on all commons is, like public access to all 'open country', a controversial and complex question.

The subject of land with public access is very complicated, both legally and historically, and is only dealt with briefly here. However, it is an important subject, as many of the paths on which conservation volunteers are asked to work are on this type of land. The problems and solutions may be quite different to those encountered on rights of way, as access to these areas is not necessarily limited to a certain line, as on a right of way.

PUBLICLY OWNED LAND

There are various categories of land owned by local authorities, statutory bodies and charities to which the public have access, either to 'wander' or on limited paths and nature trails. Statutory rights of way may of course also exist.

Country Parks

There are over 100 country parks recognised by the Countryside Commission in England and Wales, totalling about 14,000 hectares. These are mostly owned by local authorities and are managed primarily to provide informal recreation facilities. Provision for country parks was given in the Countryside (Scotland) Act 1967 and the Countryside Act 1968, with the aim of providing countryside recreation within easy reach of major urban centres, and thus to take some of the pressure off other areas of the countryside less able to withstand it. Facilities in country parks should be suitable for all ages and abilities of walkers, with emphasis on provision for families. Paths are therefore usually constructed to a higher specification than most rights of way, so they are able to cope with relatively high numbers of visitors, and use throughout the year.

Forestry Commission

The Forestry Commission is the largest landowner in Britain, owning over 800,000 hectares of plantation and mixed woodland, including large areas of heath and moorland. Since 1935, the Forestry Commission has been encouraging public access to its Forest Parks, of which there are seven in Britain, totalling 180,000 hectares. Following the Countryside Acts, the Forestry Commission has also been making provision for recreation in many other areas, with picnic sites, nature trails and forest walks, in addition to the many miles of forest road which are open to the visitor on foot.

Private and Trust land

The National Trust, a charity, is the largest private landowner in England and Wales, and allows the public free access to many hectares of moorland, mountain, downland and coastline, as well as fee-paying access to various parklands. The National Trust for Scotland owns similar types of property.

There are various other statutory and private bodies which may allow limited access to parts of their land, but recreation provision is usually secondary to the main purpose of the land management. These include some water authority land in catchments and around reservoirs, and nature reserves owned or managed by the Nature Conservancy Council, the Royal Society for the Protection of Birds, county conservation trusts and local authorities.

Routes for Recreation

Long Distance Paths

The National Parks and Access to the Countryside Act 1949 made provision for the governmental designation of long distance routes in England and Wales. The preparation and investigation of proposals for long-distance routes is the responsibility of the Countryside Commission, who also provide the Exchequer funding for all costs of setting up and managing long distance paths. Under the legislation the implementation and management of designated routes is the responsibility of the local authority, who have the complex task of negotiating all the necessary rights of way to link existing paths to form a continuous route. Thus, it was not until 1965 that the first long distance path, the Pennine Way, was opened. There are now twelve paths totalling 2528 km (1570 miles) which provide a partly new network on top of the 1949 inheritance of rights of way, and attract very much greater management effort and use. In part their success is now proving a problem, as over-use is causing erosion, particularly on high and wet moorland. The Countryside Commission, although still 'ready to consider' proposals for new routes, is now turning its attention to the development of the lower key recreational paths.

Since the passing of the Countryside (Scotland) Act 1967, the Countryside Commission for Scotland has similar powers in relation to long distance paths in Scotland, of which the first to be opened was the West Highland Way, in 1980.

Recreational paths

This is a broad category of middle distance routes, over 10 km, many of which receive grant aid from the Countryside Commission. All are described in some form of published guide. A list of the paths is available from the Countryside Commission. Although less spectacular than the long distance paths, the recreational paths are often on land more able to withstand increased visitor use, both because of the nature of the ground, and because they tend to be in wooded and cultivated landscapes that visually absorb greater numbers of people.

2 Rights and Responsibilities

There are many laws relevant to public rights of way, and this is only a brief guide. The Ramblers' Association, the Scottish Rights of Way Society Ltd, and the Commons, Open Spaces and Footpaths Preservation Society publish a comprehensive range of literature on the subject, and are very active in pursuing the legal rights of walkers. The information in this chapter does not apply to paths that are not public rights of way. Paths giving access to the various categories of land described on pages 9 and 10 are not subject to these responsibilities, unless they are also rights of way.

The following section applies to public rights of way in England Wales. Information on Scottish rights of way is given on page 14.

Definitions

The most important fact to know about public rights of way is that they are part of the Queen's highway, and are subject to the same protection in law as a trunk road.

FOOTPATH. This is a highway over which the public have right of way on foot only. It is also permitted to push a pram if the path is suitable.

BRIDLEWAY. This is a highway over which the public have a right of way on foot, on horseback, leading a horse, or riding or pushing a bicycle.

BYWAY OPEN TO ALL TRAFFIC. This is a highway over which the public have a right of way for vehicular and all other kinds of traffic, but which is used mainly for the purpose for which footpaths and bridleways are used. In some counties such tracks are still referred to as Roads Used as Public Paths (RUPPs).

STATUTORY LAW. This is written law, made by a legislative authority. Statutory rights of way are those whose status was confirmed following the National Parks and Access to the Countryside Act 1949, or those created by an Order (see below). Most public rights of way in England and Wales are now statutory rights, except for those whose status has not yet been decided.

COMMON LAW. This is unwritten or customary law, based on judicial decisions. Common law rights of way are those used by custom, although in the case of dispute, the existence of this common right may have to be decided in a court of law. Prior to 1949, all rights of way in England and Wales were of this nature. Many rights of way in Scotland are common law rights of way.

ORDERS. Orders are made by County Councils, District Councils or the Secretary of State for the Environment, for the creation, diversion or extinguishment of public rights of way. They usually arise through the request of the landowner. Extinguishments can be made if the local authority considers the path is no longer needed for public use. The grounds for diversion are that it is in the interests of the owner, lessee or occupier, or of the public, and that it is not substantially less convenient to the public. The effect on public enjoyment of the path as a whole must also be considered.

DEFINITIVE MAPS. The definitive map for each county is the result of the surveying and recording of public paths required under the National Parks and Access to the Countryside Act 1949. It is conclusive evidence in law as to the course of a public right of way. The status shown does not preclude the existence of higher rights. The information is shown on a map to the scale of 1:25,000 or 1:10,560 and descriptions of the paths may also be given in an accompanying statement. The local definitive map is available for consultation by the public at County Council and usually District Council offices. The map is updated to show the results of confirmed Orders, or changes in a path's status or existence resulting from documentary or public evidence. The information is included on Ordnance Survey maps as these are updated.

Ownership

Footpaths, bridleways and byways are said in law to have been 'dedicated' to the use of the public by the owner of the land they cross. Very few have been formally dedicated, most being 'presumed dedicated' because the public have used them uninterruptedly for at least 20 years. Any surface on the path belongs to the County Council, but the soil beneath it is the property of the landowner.

MANAGEMENT AND MAINTENANCE

The local authority ultimately responsible for 'asserting and protecting the rights of the public to the use and enjoyment' (Highways Act 1980) of paths in their area is the County Council. District Councils, Parish Councils, Community Councils (the equivalent of Parish Councils in Wales) and National Park Authorities may take over some of the responsibilities outlined below. Contact the County or District Council office to find out who is responsible in the area with which you are concerned.

11

Voluntary groups who want to work on public rights of way must arrange to do so with the relevant authority, who is then responsible for dealing with any claims for damages which may arise. The landowner must also be consulted either by the authority or directly by the group. The procedure depends on the system, if any, operated by the particular local authority. A few local authorities do all maintenance work necessary, and no voluntary help is needed; others neither maintain the paths efficiently nor encourage voluntary effort. Most however, are pleased to involve volunteers, and may give assistance with materials, tools and transport.

Surface

The local authority is responsible for the maintenance of any surface, such as tarmac, gravel or chippings, overlaying the soil of the path.

There is a statutory right to plough any footpath or bridleway which crosses a field. Footpaths and bridleways along the edges of fields and all byways open to all traffic never have a right to be ploughed. Prosecution can be brought by a local authority against anyone who ploughs a footpath or bridleway without a right to do so.

If a right of way is ploughed, the landowner must restore the surface within two weeks of the date on which ploughing began. In exceptional weather conditions, this period can be extended to a date 'as soon as practicable thereafter'. Restoration is usually taken to mean a pass back and forth with the tractor to mark the route, and discourage crop growth. A better walking surface is made if the route is rolled.

Growth

The local authority is responsible for clearing any growth, except a crop, from the surface of the path. This includes, for example, nettles or bracken which grow in the line of the path and block it. Any overhanging growth from trees, hedges or shrubs adjoining the path is the responsibility of the landowner. The local authority can require the landowner to carry out this work, and if necessary, can do it themselves and recover costs. Overhanging herbaceous growth such as grasses and nettles is in theory not a problem, as the walker can still pass along the path. In practice, this often is a problem, which if dealt with at all, is done by volunteers or by the more efficient local authorities.

A standing crop is the responsibility of the landowner, as this results from failure to restore the surface after ploughing. As the local

authority can restore a ploughed surface and recover costs if the landowner fails to do so, this could involve clearing a standing crop. A standing crop which is tall enough or of a nature to block the path is treated in law as an obstruction.

Stiles and gates

Stiles and gates on footpaths must be maintained by the owner in a safe condition, and to a standard which does not make an unreasonable interference to the rights of users. A minimum 25% grant must be contributed by the local authority towards the cost of erection or maintenance of a stile or gate if the owner requests it. An owner wishing to erect a stile or gate on a right of way where none existed before must apply to the County Council for consent.

In practice this system usually leads to a poor standard of stiles and gates, as many owners either ignore them or do minimum repairs, and rarely apply for grants. Some local authorities prefer to supply and erect stiles themselves, in order to get the work done to a proper standard. Many voluntary groups also erect stiles, usually supplied by the local authority.

Bridges

The local authority is responsible for the maintenance of bridges on rights of way. Bridges that cross canals or railways are maintainable by the canal authority or British Rail.

Signposting

The local authority is required to signpost all junctions of footpaths, bridleways and byways with metalled roads. No time limit was given for this work to be done, and many counties have not yet completed this task.

Access for maintenance

The regulations prohibiting vehicles on footpaths and bridleways do not apply if the local authority (or the group acting as agent for it) need access by vehicle for maintenance or improvements to be carried out. The local authority is liable if any damage to the landowner's property occurs.

Obstructions

Obstructions can include a barbed wire fence, a heap of manure or a standing crop. The local authority can order the landowner to remove the obstruction, or remove it themselves and recover costs. A member of the public is allowed to remove as much of an obstruction as is necessary to get past it, but must be a bona fide traveller, and not have gone out with the specific aim of removing the obstruction.

Bulls

Under the Wildlife and Countryside Act 1981, an occupier of a field or enclosure crossed by a footpath, bridleway or byway can be fined up to £200 for the following offences:

a   Having any bull over 10 months of age in a field unaccompanied by cows or heifers.

b   Having a dairy bull over 10 months of age in a field with heifers or cows. Dairy breeds are Ayrshire, British Fresian, British Holstein, Dairy Shorthorn, Jersey, Guernsey and Kerry.

Thus, non-dairy breeds over 10 months accompanied by cows or heifers can be at large in fields or enclosures crossed by public paths. These regulations do not apply to open hill areas.

Use

It is a statutory offence to drive a vehicle along a bridleway in the absence of a private right to do so. The general prohibition on driving off roads does not apply if the vehicle is within 15 yards of the road, and is driven there only for the purpose of parking. However, the landowner can order the vehicle off even within this limit, and can sue for trespass.

It is not a statutory offence to ride a horse or a bicycle along a footpath, unless there is a bye-law or a traffic regulation order prohibiting that activity, but a person doing so without the permission of the landowner commits a civil trespass, and the landowner can order them off, and where appropriate, sue for damages.

As public rights of way are by definition highways, they can only be used for bona fide journeys. The traveller can do anything reasonably ancillary to the journey, such as pausing to look at the map, or sitting down for a rest, but any activity not directly referable to the right of passage can be an act of trespass. One example of when a landowner successfully sued for trespass was in 1900, against a journalist who walked up and down a right of way taking notes on the performance of racehorses training on the adjoining ground. Although this may seem trivial, it does underline

the limit to the public's use of a right of way. It is, as the name describes, only the right of passage.

Trespass

Trespass is the unauthorised entry of a property, and is a civil wrong, not a crime. Although legal action is unlikely, a trespasser can be sued for this alone. The owner would only get nominal damages, but the trespasser may be held liable for legal costs which could be expensive. If it could be shown that damage was done to the property, the trespasser would be liable for the payment of compensation. Trespass on railway land is a criminal offence.

Carrying of tools

It is usually considered permissible for a walker to carry, as part of his right of passage, light tools such as secateurs for cutting back odd pieces of overhanging growth. However, the carrying of any heavy tools such as a saw might be considered evidence that the person was not a bona fide traveller, but was out on purpose to remove an obstruction. The removal of such an obstruction might constitute damage of property.

Organised groups must get the relevant permission to do work as described on page 12. It is always best for work to be done with the approval of all parties, and although 'do-it-yourself' maintenance may seem to be the simplest way of getting work done, it could lead to antagonism between the landowner and the public.

Permissive Paths

These are not public rights of way, but footpaths or bridleways used by public with the permission of the landowner. They should have a notice to that effect displayed on the path to make it clear that the owner does not wish to dedicate the path as a right of way, and the path cannot then be 'presumed dedicated' after 20 years use by the public. The owner can register with the Highway Authority that he has no intention to dedicate. Most of these paths are in National Parks or other areas where recreational use is intensive, and any necessary maintenance of permissive paths is done by the local authority and not by the landowner. The path can be closed at any time by the landowner.

Although such paths do not have the legal status of a right of way, and are not acceptable as a substitute, they do allow access to be opened up by a simple procedure that is usually more acceptable to the landowner than a creation order, which gives the public right of way for all time. See also paragraph (c) on page 168.

Licensed paths

These are similar to permissive paths, and have been used particularly in East Sussex to link rights of way to form a continuous footpath or bridleway. The council does all work of making and maintaining access, and must restore the land to its original condition if the agreement is terminated, which can be done at three months' notice by either party. Like permissive paths, these allow improvements to be made to the path system with the minimum of red tape.

Scotland

There are several important differences between English and Scottish laws on access and rights of way.

Trespass

Any person entering onto land without the owner's permission is trespassing. In Scotland simple trespass is neither a civil nor a criminal act, but an aggrieved owner may take action in the courts by interdict if he can prove damage is being caused. In such instances the courts need to be satisfied that the trespass will be repeated by the same offender and that there has been some appreciable loss or inconvenience.

Public rights of way

There is no Scottish equivalent to the 1949 legislation of England and Wales. Public rights of way in Scotland, be they footpaths or bridleways, are of two types:

1   A right of way whose existence can only be established in common law by meeting the following four requirements:

   a It must have been used by the general public for a continuous period of not less than 20 years.

   b The use must be as of right, and not attributable to tolerance on the part of the proprietor.

   c It must connect two places to which the public habitually and legitimately resort.

d It must follow a route more or less defined.

A common law right of way can be lost if it is shown that the public have not used the path for 20 years. This disuse must be by choice, and not because the owner takes action to discourage use.

These are in many ways similar to the common law criteria adopted by National Parks and Access to the Countryside Act 1949, for identifying public rights of way for inclusion on Definitive Maps in England and Wales. In Scotland, common law is still the basis of law on public rights of way, and there are no statutory requirements on local authorities to prepare such maps.

2   Statutory rights of way, which are those created by an agreement or an order. Agreements with landowners can be made by District or Regional Councils, or if agreement is impracticable, an order can be made. Orders for creations or diversions do not take effect until they are confirmed by the Secretary of State, and there is a procedure for objections to be heard.

Statutory rights of way are either footpaths or bridleways. Bridleways are defined in similar terms to those of England and Wales, but footpaths can be used on foot and on a bicycle.

Maps

In 1981 the Countryside (Scotland) Act 1967 was amended to provide that duties connected with public rights of way in Scotland should be the responsibility of the general or district planning authority for any particular area. As already mentioned, these responsibilities do not extend to preparing Definitive Maps, but many councils have prepared their own maps of claimed public rights of way in their areas, often with the assistance of community councils. The only statutory requirement on authorities is that they must have maps which the public can consult showing land subject to access agreements or orders acquired under Section 24 or 25 of the 1967 Act. The Scottish Rights of Way Society maintain maps of the more major public rights of way at a scale of 1:50,000 and the Countryside Commission for Scotland have made these maps available to all planning authorities in Scotland.

Maintenance

Local authorities must carry out sufficient work to bring a statutory right of way into fit condition for public use, and to maintain it in that condition. Their responsibility for common law rights of way are not so positively stated, and extend only to 'asserting, protecting, keeping open and free from obstruction or encroachment'. As in England and Wales, many rights of way are not maintained, although in Scotland this is not always a breach of statutory duty as it is in England and Wales.

Obstructions and bulls

The ruling on bulls is similar to that in English law.

A member of the public may, without prior notice or protest, remove as much of an obstruction as is necessary in order to restore free passage. The owner may plough a public right of way, unless excluded from doing so by the terms of an agreement or order. The local authority must be informed by the owner of his action within the seven days following ploughing, and the owner must reinstate the surface as soon as possible.

Other access

There are similar provisions to those in England and Wales for the making of access agreements to open land, and the establishment of country parks.

Voluntary action

As claimed public rights of way under common law can be closed if not used for 20 years, their use and maintenance is of importance to walkers. Limited clearance, sufficient to restore reasonable passage, can be done by members of the public without informing either landowner or local authority, but it is obviously best for any work to be done with the approval of both landowner and local authority.

Care needs to be taken to ensure that such work is limited to this end, to avoid giving rise to charges of damage from a landowner or occupier. In these circumstances, it is obviously important to liaise with the planning authority and the owner and occupier of any land in question before doing anything more than this. If difficulty is encountered in identifying who to approach other than the planning authority, contact should be made with the local representatives of the National Farmers' Union of Scotland and the Scottish Landowners' Federation.

3 Path Planning: topography and habitat

This chapter is concerned with the physical limitations of soils, slopes and vegetation for paths, and Chapter 4 looks in more detail at the way people use paths. Because of the infinite variety of sites and problems that occur, it is not possible to provide any easy-to-follow formulas, but only to suggest a process by which decisions can be made.

Why Survey?

Even for an apparently straightforward task, it is worth gathering as much information as possible about the site.

a   A physical survey is essential. This may vary from mapping quite a large area to choose a new route, down to checking that the position for a signpost hasn't a large boulder just beneath the surface.

b   Unless you know the site very well, it is difficult to say what it will be like under different weather conditions and other times of year. A careful survey should reveal some clues.

c   The very process of making a survey, taking notes and photographs gives the opportunity for the problem to be thought over, and hopefully for the best solution to emerge.

d   Written notes and drawings are essential for the task leader, as well as for the person who plans the work. They are a useful record for the future.

e   An estimation of quantities of materials, labour and time required can only be made if there is reliable information on which to base it.

f   A survey should give some idea as to who uses the path now, and who may use it in the future. This is the most problematical part of path planning, and is affected by many factors outside the scope of this book.

g   It is easy to get carried away with a favourite project, or to simply follow plans made at an earlier date which may no longer be relevant. It is better at the outset to consider the possibility of not beginning the project, than to abandon it half way through.

The main site features to be considered are:

1   Soil

2   Drainage
3   Slopes
4   Vegetation/habitat
5   Artifacts
6   Levels of use
7   Feasibility

The advice in this chapter mainly relates to rights of way in open country, and paths that are not rights of way, where a choice of the exact line can be made. However, similar procedures can also shed light on physical problems on rights of way, even where the route is not open to choice.

Recording the Information

Maps

It is easiest if you have a map to work from, preferably the Ordnance Survey 1:10,560, enlarged to 1:100, 1:200 or 1:500. The local authority may be able to supply a copy. For larger schemes, the 1:25,000 may be sufficient. The 1:10,560 does not show contours, but it is quite easy to transfer the contours from a 1:25,000 map with the aid of an overhead projector. Trace the contours off the 1:25,000, and project the tracing onto the enlarged 1:10,560 mounted on a wall or board.

If you do not have a map, then make one. You will probably anyway need a sketch map to record various features, and as long as it is made clear that the sketch map is just that, and not drawn to scale, the execution is not too important. It is too easy to look at an area and think you will remember it, when a quick sketch map will make you observe more carefully whilst in the field, as well as being useful for reference and drawing plans later.

Sites for bridges and steps need careful measurement, and details are given in the relevant chapters.

Photographs

Photos are useful not only in planning the work, but for later assessing its effectiveness. For photos to be useful for future work, it is essential to be systematic in taking them. A single lens reflex camera is preferable, using black and white print or colour transparency film. A wide angle lens is useful.

1   Record the site, date, film type, camera type and lens focal length.

2   Write a description of each point from which

16

you take a photo, so that the point can be found again. It may also be possible to mark them on a sketch map. Photos may be combined with some sort of marking system on the ground, such as transects (see p29). It is useful to have a conspicuous marker in the photo, because it enables you to easily identify the photo and match it with your notes. Beware of taking a whole series of photos, with a series of carefully written notes, and then being unable to match the two.

Markers can be a problem, as inconspicuous ones are difficult to re-locate, and conspicuous ones attract attention and may be removed. The choice will depend on the type of vegetation, and the use of the site by people and stock. In general, well-made conspicuous markers are better, as these look 'official' and tend to be respected by walkers at least, if not by stock.

Soil

Soil consists of mineral particles, organic matter, air and water, and is present in layers overlaying the bedrock. The layers form the 'soil profile' (see examples below). The depth, structure and composition of the soil, and its permeability, are all significant in its ability to withstand wear on paths. It is useful to look at the soil both for planning a new path or diversion, and for carrying out remedial work on an existing path.

Soil on paths may be affected in two ways outlined below, both of which may lead to soil being washed away from slopes. This process, called erosion, can leave deep gullies or exposed bedrock, which are uncomfortable to walk on, unattractive, and unable to support vegetation.

a   Compaction. Trampling compresses the mineral and organic matter in the soil with resultant loss of pore space. This lessens the soil's ability to absorb water, and to support plant life. Rain water falling on compacted soil will lie as puddles on flat ground, or will flow down slopes, causing erosion.

b   Churning. The action of feet and hooves on bare ground loosens the surface particles, which are then easily carried away by water or blown by wind.

TESTING SOILS

The vegetation and appearance of the ground will give a good idea of the type of soil beneath, and it may then only be necessary to do a few tests by way of confirmation. If a completely new path is being planned, particularly in wooded country which is not easy to assess quickly, a more detailed survey is worth doing.

Methods of testing the soil include the following:

a   The quickest way to test soils is to use a soil auger (p181 3.1). This simply screws into the ground, and when removed, brings out a sample of soil. Extra lengths are added as required. The depth and composition of the soil can thus quickly be checked.

Always mark the location of the test sample on the sketch map, and note down your observations. You may find it useful to keep some samples in polythene bags, so you can compare them as you go along, and thus make some judgements even if you are not sure what the material is. Otherwise, without the skill to describe a sample accurately, you will find by the 4th hole or so that you can't really say how it differed, if at all, from the first. Remember to put a slip of paper into the bag noting the location of the sample. Beware though, of just collecting samples and being unable to make sense of them later. As with all survey work, it is much better to make a few simple observations and come to some common-sense conclusions while you are in the field, than to just 'collect data' and think you will interpret it later, because probably you never will.

b   The best method is to dig a test hole with a spade. This is harder work than using an auger, but exposes the soil profile and makes the soil easier for a beginner to assess. Try to dig the hole as neatly as possible, both for economy of effort, and to expose a clear profile. Always fill in the hole and tidy up afterwards.

c   A field test used in engineering to test the strength of soils measures the result of applying a standard weight to the soil. The strength of the soil is its ability to support loads, and depends on the type of soil, its moisture content, and state of compaction. A similar, if less accurate test can be made by counting the number of blows needed to knock a post into the ground, trying to use an equal force for each blow. This can be used for planning boardwalks or surfaced paths across soft or wet ground, where you need to test the depth and the strength of the soil.

There is no point in designing a boardwalk supported on posts which are subsequently found to disappear without trace into deep peat.

d   The simplest soil tests involve checking that a planned scheme is practical for a certain site. Examples include ensuring that sites for stiles and signposts have enough depth of soil or excavatable material to secure the posts. This is especially important for flights of steps, which are frequently on the thin soils found on slopes. Make sure that the chosen route has deep enough soil to give secure anchorage for the steps. If no such route can be found, other methods of anchorage will be necessary (see Chapter 10).

Drainage

This is the most critical factor in determining the suitability of soils for paths.

Excess water is indicated by any of the following:

Surface water

Water lying on the surface, other than immediately after heavy rain. Previous waterlogging is indicated by the indented, damaged surface caused by the trampling of stock on wet ground, called poaching. Look at the colour of the soil (see below) to determine whether waterlogging is permanent, seasonal or merely temporary.

Vegetation

Plants that indicate wet conditions are described in the section beginning on page 24.

Soil colour

A well drained soil is a uniform brown or reddish-brown in colour. Excess water at the surface or in a test pit of this colour soil is likely to be caused by a localized drainage flow, which can be remedied by intercepting the flow at source or away from the path. These soils have a good structure and are suitable for paths.

A soil subject to permanent waterlogging becomes a fairly uniform greyish, bluish or blackish colour. This is due to the water in the soil being stagnant and excluding air, thus causing the reduction of ferrous oxide to grey ferrous iron. This process

is called gleying, and can occur in many different types of soil. The colour can vary from a dense grey in clay soils, to a greyish tinge in peaty or sandy soils. This problem is likely to be difficult to cure, as it usually results from general waterlogging over a wide area, and there is no identifiable source of water which can be diverted. These soils are likely to become worse drained if subject to trampling, as further compaction and loss of pore space will occur. A path over such soil should have a surface which raises it above the waterlogging, with side drains as necessary. Sometimes gleying is due to the presence of a hardpan, which can be remedied (see below).

A soil which is seasonally waterlogged has a mottled appearance, with patches of a rusty-brown colour where the seasonally increased air content has caused the oxidation of ferrous iron to ferric oxide. In general these soils are better avoided, but it may be possible to remedy the problem by putting in a network of drains which collect the water and take it away from the path. The removal of water allows more air into the soil, which aids growth of vegetation. This sets off a chain reaction which ultimately improves the soil; the increased plant growth taking up further water from the soil, more roots increasing fissures and air space in the soil, and thus helping to 'dry out' waterlogged ground.

Root development

A well drained soil has roots growing deep into it. Roots of most plants are killed in waterlogged soil, and the lower limit of root growth will indicate the level of waterlogging.

Earthworms are also a good indication of soil wetness. They can live neither in dry nor water-logged soils, therefore their presence indicates a reasonably drained but cohesive soil, which should bear trampling and be resistant to erosion.

Having identified the presence of excess water from any of the above features, its cause should be considered. This is usually determined by looking at the topography, and the pattern of surface drainage.

CAUSES OF EXCESS WATER

High water table

This is the level of the water in the ground, which may lie on or just below the surface, or emerge as springs or seepages on slopes. The source cannot be localised, as it includes rainwater falling over a wide area, and water rising from deep underground sources. It usually causes peaty or gleyed soils, with typical moor or wetland vegetation. There is little that can be done about it, except for major drainage schemes beyond the scope of path management. Springs or seepages above or on paths can be intercepted, and drained across or away from the path.

Lack of permeability

Lack of permeability in the soil results in rain or surface water from streams being unable to drain downwards through the soil. This can be improved by laying drains.

Clay soils

The permeability of clay soils can be increased by adding gravel or large amounts of organic matter, but this is not practical on paths. The usual solution is to make a camber or crossfall at the surface, so that water runs off the path and is drained away in a side drain.

Ironpan

This results from the movement of minerals downwards through the soil profile, a process called podzolisation. It can occur on quickly draining sandy soils, as well as on upland areas with high rainfall and low evaporation. The minerals, mainly iron and manganese, are re-deposited in a layer which may become imperm-eable, called the ironpan. The soil profile will show a bleached and often gleyed, waterlogged upper layer, and the reddish-brown cemented ironpan below. The ironpan may be up to 100mm thick, or it may be very thin, resembling a walnut shell. The soil beneath the ironpan may be quite free draining.

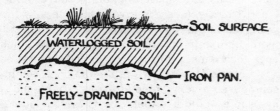

If the ironpan is found to be only about 300mm below the surface, the top layer can be stripped, the ironpan broken up, and the path surfaced with free draining material. Alternatively, the breaking up of the ironpan and the mixing of the top layer of soil will improve permeability.

Ironpan and gleyed soils are frequently associated with peat, which builds up in the anaerobic conditions. These conditions inhibit the activity of bacteria which break down plant remains. Podzolisation and the formation of ironpan is unlikely to occur beneath broad leaved woodland, as the deep-rooted trees circulate water and dissolved minerals up through the soil.

Peat

Peat is very difficult to deal with as it is formed over impermeable soils, and like a sponge can continue taking up water until it is super-saturated. If artificially drained with ditches it may dry out irreversibly, and be liable to wind blow. If the peat is less than about 300mm deep, it can be stripped along the line of the path, and a free draining surface laid over the impermeable soil beneath. Deep peat can only be dealt with by building a 'floating' path, either as a boardwalk, or loose surfacing over a membrane such as Terram (see p78).

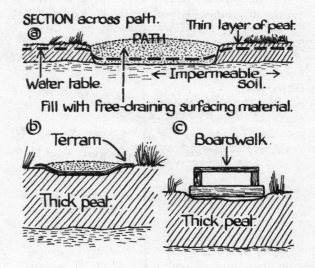

Slopes

The following methods can be used for measuring the gradient or angle of a slope.

a   Ordnance Survey maps are only useful for large schemes, because even the most detailed information, on the 1:25,000 map only shows the contour interval every 25 feet. It does help with understanding the general topography and drainage patterns of an area, but is not often of use for the detailed route finding that may be necessary.

b   The most accurate form of measurement is made by using surveyors' levelling equipment. It may be possible to arrange for the local authority or a field study group to do such a survey, but this degree of accuracy is only occasionally needed for designing bridges and steps. See the relevant chapters for details.

c   The quickest and easiest method of measuring angles is to use a clinometer. This is a small, hand-held instrument, from which the angle is read off in degrees. Clinometers are robust instruments and are very suitable for this type of work (p181 3.2).

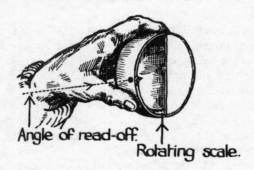

Angle of read-off.   Rotating scale.

A cheap but effective clinometer can be made with a spirit level, straight pole and plastic protractor. Sight along the pole, and tilt it until it is parallel with the slope. Then measure the angle between the pole and the spirit level, checking the latter is horizontal.

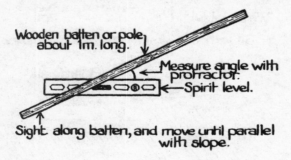

Wooden batten or pole, about 1m. long.

Measure angle with protractor.
Spirit level.

Sight along batten, and move until parallel with slope.

Frequently tasks involve re-routing an eroded path up a slope where most, if not all, of the slope is visible from the bottom. A new route can be chosen by finding the appropriate gradient (see below) with the clinometer, and then marking it on the ground with pegs. Even if the choice of line appears obvious, it is always worth walking the slope and looking at it thoroughly, not only for the gradient, but for soil depth, wetness, vegetation and lines of sight. It is particularly important to look at the slope from the top downwards, as it is the downhill use that causes most of the problems (see p36).

The optimum gradient for a stable path is one that drains quickly without causing erosion. The

diagram below indicates the range of different gradients that affect path construction and repair. These gradients apply to the gradient of the slope itself, not the path. A path taking an oblique line up a slope will have a gradient less than that of the slope.

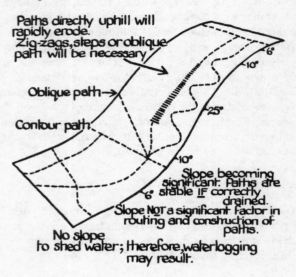

Paths directly uphill will rapidly erode. Zig-zags, steps or oblique path will be necessary

Oblique path →

Contour path

Slope becoming significant. Paths are stable IF correctly drained.
Slope NOT a significant factor in routing and construction of paths.

No slope to shed water; therefore, waterlogging may result.

Problems on slopes

Erosion becomes more damaging as a slope steepens, because the erosive power of water increases with increasing speed of flow. As it is a natural process which occurs even on vegetated slopes, soils tend to get thinner as the slope becomes steeper.

The likelihood of damage to slopes increases when vegetation cover decreases. Trees not only break the impact of rainwater, but make the management of people on paths much easier as lines of sight and movement are restricted. Most hillsides in Britain are not wooded, and it is their open nature which is especially cherished by walkers.

GRADIENT

The following diagram shows different gradients with corresponding characteristics and uses. Note the difference between gradients as they apply to slopes, and as they apply to paths. Conversions to % and 1 in x values are given as these may be encountered in other literature. Note also that the actual slope in the field always appears much steeper than does a diagrammatic representation. The 30 degree slope shown in the diagram would be considered a very steep slope to climb, and many people would use their hands for assistance in climbing it. The Idwal Slabs in North Wales, which appear so impressive when swarming with rock climbers, are in fact only about 40 degrees.

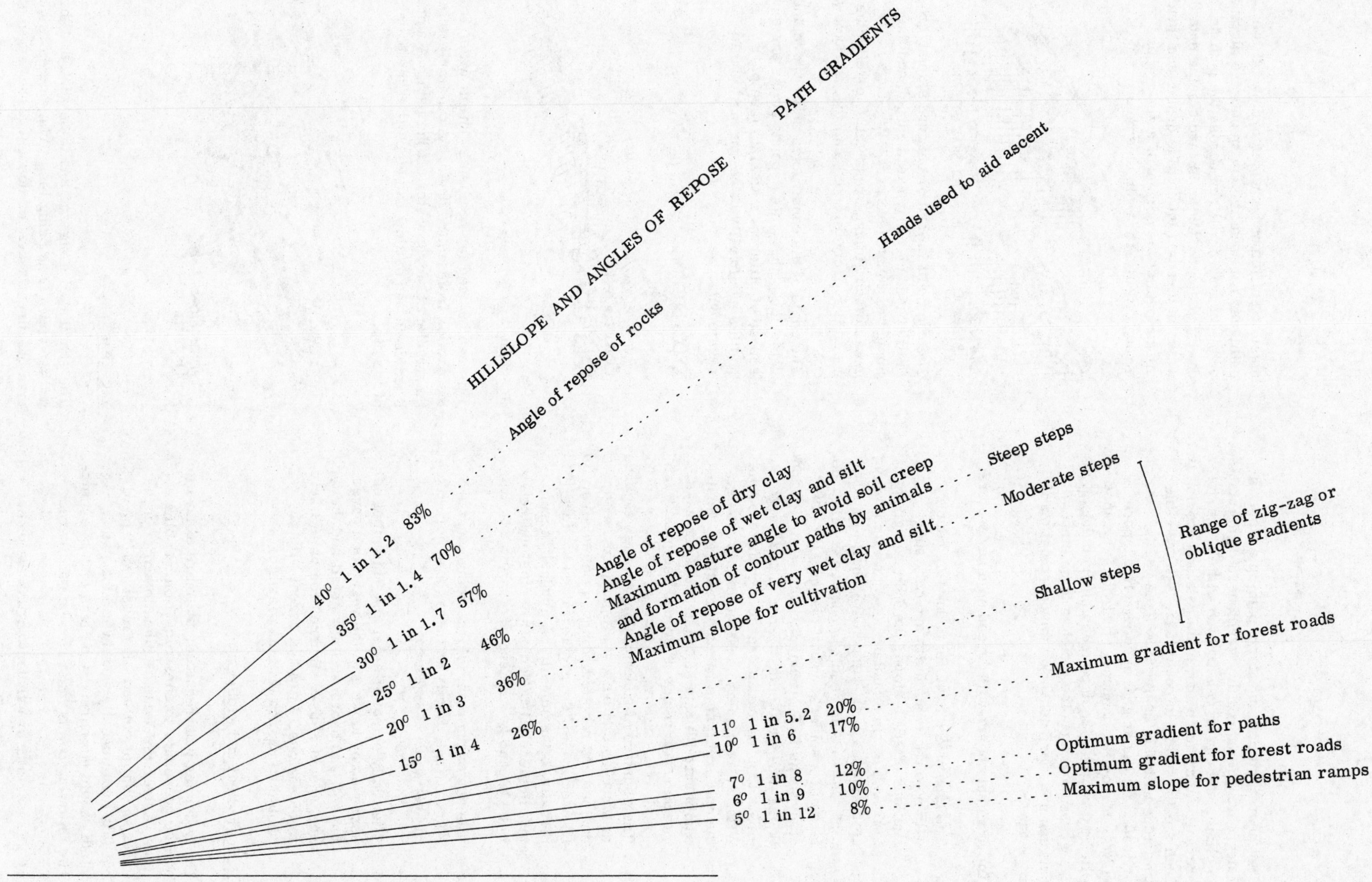

HILLSLOPE AND ANGLES OF REPOSE

PATH GRADIENTS

Hands used to aid ascent

Angle of repose of rocks

Range of zig-zag or oblique gradients

Steep steps

Moderate steps

Shallow steps

Maximum gradient for forest roads

Optimum gradient for paths

Optimum gradient for forest roads

Maximum slope for pedestrian ramps

Angle of repose of dry clay and silt

Angle of repose of wet clay to avoid paths by animals

Angle of repose of contour wet clay and silt

Maximum formation of very wet clay

Maximum formation and repose for cultivation

Angle of repose for cultivation

Maximum slope

40°	1 in 1.2	83%
35°	1 in 1.4	70%
30°	1 in 1.7	57%
25°	1 in 2	46%
	1 in 3	36%
20°	1 in 4	26%
15°		
11°	1 in 5.2	20%
10°	1 in 6	17%
7°	1 in 8	12%
6°	1 in 9	10%
5°	1 in 12	8%

The theoretical optimum for a path up a slope is 7 degrees. A path of this gradient is comfortable to climb, and can be constructed with drains to prevent water running down the path and eroding it. Such a path is easy to route if the 7 degree line coincides with the 'desire line' (see p32) or leads to a linear goal such as a flat topped ridge. More often though, the obvious destination will be directly up a slope much steeper than 7 degrees, and a choice then has to be made between a steeper ascent or the construction of zig-zags.

The gradient chosen will depend on many factors:

a   The type of country. In undulating country with no obvious summits and only short sections of steep slope, the 7 degree gradient is possible. In mountainous country with long steep slopes such a gradient is quite impractical, as it would result in each peak being the focus of a spider's web of zig-zagging paths. Many paths in Scotland, the Lake District and Wales have gradients steeper than 25 degrees, most of which are subject to serious erosion. With use remaining constant or increasing, the only solution is to make gentle gradients combined with sets of constructed zig-zags, together with repair of eroded sections.

b   The standard of construction. If the resources are available to build a properly surfaced, drained and revetted path, it is more likely to be used even if the gradient does not lead it directly to the summit. A gradual ascent that is merely trodden out is most unlikely to be kept to by walkers.

c   The use of the path. Horseriders, and to some extent, family walkers, are more likely to keep to an easy ascent.

d   The likely extent of erosion. This must be assessed by looking at the vegetation cover, the soil and bedrock, the rainfall and the type and amount of use. Where erosion is likely to be serious, a gradual ascent with zig-zags should be chosen if at all possible.

Path gradient and width

The wider the path required, the more excavation will be necessary to construct it on an oblique line. Calculations can be made to work out the exact amount of cut and fill (Vogel, 1971 pp 61-64) but this is not usually worthwhile for paths. The width should not normally be less than one metre. The following points must also be considered:

a   Cross-fall, or gradient across the path. This

must be sufficient for water to flow off the path, and not down it. On an outward sloping path, the water will run across the path and down the slope. On an inward sloping path the water is directed into a side drain, and then off the path at the corner of a zig-zag or through a culvert or dip (see below).

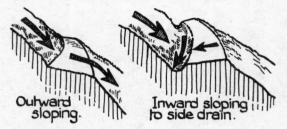

Outward sloping.   Inward sloping to side drain.

The choice of inward or outward sloping cross-fall depends on the type of bedrock and the amount of runoff (see p75). Zig-zags must have inward sloping cross-falls, or else the water is merely directed onto the path below.

b   The angle of the constructed slope above and below the path. This depends on the angle of repose of different materials (see p21) and the availability of resources to build revetments.

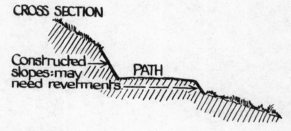

CROSS SECTION

Constructed slopes: may need revetments.   PATH

Gradient dip

This can be used for removing run-off water from the path or from a side drain. A dip is made in the gradient to force water off the path. A stone-lined drain and outflow can be built at the bottom of the dip.

Water flows off path.

Zig-zags

The gradient of a zig-zag has to be such that the walker views it as being less tiring and so quicker than the direct ascent or descent. The length of the limb of the zig-zag is also significant. Repetitive limbs of the same length, such as on the zig-zag path up Selborne Hanger, Hampshire, are tedious to walk, and two obvious short cuts develop down either side. Drainage is also a problem, as each limb directs water down the short cut.

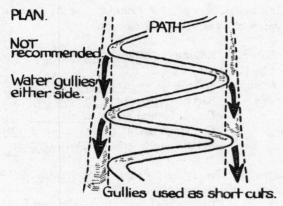

PLAN.

NOT recommended.

Water gullies either side.

PATH

Gullies used as short cuts.

On the other hand, too long a limb may appear to lead away from the direction of the destination. Varying the length of limb is the best pattern, taking advantage of any spurs, outcrops, boulders or slight lessening of gradient to make the turn. The recently constructed zig-zag at Nab Scar, Cumbria, takes a very natural and comfortable line up the slope.

Ideally, the corner of the next zig-zag above should not be visible, but this is not often possible. The zig-zags below are nearly always visible, except in tall vegetation, and short cutting has to be discouraged by steep revetments or barriers.

A study has been made of some of the zig-zags on the stalking tracks in the Mamores, Highland Region (John MacKay; unpublished paper). These tracks were constructed during the 19th Century to give access for shooting parties, and have long gentle gradients alternating with spectacular sets of zig-zags to gain height. The zig-zags often follow the nose of a ridge, where run-off from the mountain is less than in a gully.

The histograms below show the limb gradient and length of limb for 48 different sections of zig-zag.

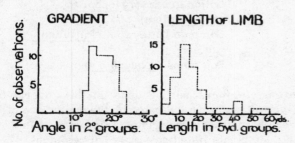

These zig-zags are possibly amongst the steepest

in Britain, climbing slopes of over 25 degrees. The histogram therefore shows a higher mean gradient than would normally be found, limb gradients of 15 to 16 degrees being more typical.

Another example of zig-zags is the pony track to Ben Nevis, which climbs a slope of 23 degrees, with zig-zag limbs of maximum 11 degrees. However, in this situation the gradient does not give a direct enough route for most walkers, and the zig-zags have been largely abandoned in favour of a heavily eroded direct route up Ben Nevis.

The corner of a zig-zag is the point most vulnerable to erosion, and should be positioned on a stable part of the slope. This may be an area of bedrock exposed on a scree slope, or above an outcrop or bluff on a vegetated slope. It should also be positioned at a point where the slope is less steep, to allow more room for the turn to be made. The turning area should only have a very slight gradient, as shown in the diagrams below. If the gradient of the limb is increased just below and above the turn, this increases the difference in height between points A and B, and discourages short cutting. Revetments are normally needed to prevent the constructed slopes from collapsing.

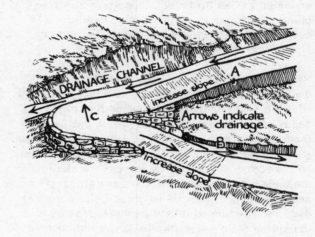

Alternatively, the turn can be made with steps.

Habitat

This section is particularly relevant to areas where there are few constraints of topography or land-use which dictate the route of a path, and the route can be chosen according to variations in the habitat. The type of vegetation is a good indicator of drainage patterns and seasonal water-logging. Vegetation is also significant because of its varying resistance to trampling, and its use for screens and barriers.

HEATH AND MOOR

These habitats are explained in some detail as it is on such areas that the majority of 'route planning' work is done. They are also the most critical, being physically the least suitable, but aesthetically and geographically amongst the most popular areas for walking.

The types of habitats listed below are not separate, and gradations exist between them. There are few trample resistant plants such as the grasses, and removal of the vegetation can result in rapid erosion of the peat and soil. All these habitats have a low fertility because of the nature of the underlying rock or gravel deposits. The range of plants which can grow in the acid conditions is relatively low, and pioneer vegetation is very slow to establish on bare areas.

Where poor drainage or high rainfall occurs, bacterial activity is inhibited, and this, together with the tough nature of the vegetation slows the decay of dead plant material, which builds up to form peat. Deep peat acts like a sponge, and is a very poor walking surface. The problem is not so severe on dry heath, as the peat is shallow and the underlying material is free draining. On moor and bog the peat may be anything up to 5 metres deep, and is turned into a soggy morass by trampling once vegetation is destroyed. Under heavy use the damage quickly spreads laterally as walkers try to avoid it. Various methods are given in this handbook to ameliorate such problems, but the best solution is to re-route paths off deep peat.

Heathland

This develops on light sandy soils in areas of low rainfall, mostly in southern England. The dominant vegetation is heather (Calluna vulgaris), but this is replaced in damp areas over hardpan or pockets of clay by the species listed below.

A typical sequence on dry heath is that trampling destroys the heather which exposes the thin peat layer to erosion by rain or wind, leaving a sandy or gravelly path. This is satisfactory on flat areas, and gives a dry path which should not spread laterally unless use greatly increases. The pale scar of the path may mar the appearance of the heath. On steep slopes serious gullying can occur as sand and gravel are washed away. This has occurred, for example, at the Devil's Jumps at Frensham in Surrey.

Dominant	Heather (Calluna vulgaris)
Wetter areas	Sedges (Carex spp) Sundew (Drosera rotundifolia) Cross leaved heath (Erica tetralix) Cotton grass (Eriophorum angustifolium) Purple moor grass (Molinia caerulea)
Drier areas	Wavy hair grass (Deschampsia flexuosa) Bird's foot trefoil (Lotus corniculatus)

Calluna moor

This develops on ancient impermeable rocks in areas of high rainfall, mainly in the north and west. The soil is shallow and poorly drained. The dominant plant is still heather, in spite of the waterlogged conditions being very different from the dry sandy heaths described above. The reason is that the cold waterlogged soils become oxygen deficient, and this together with strong winds makes the uptake of water difficult, and plants are in physiological drought. Heather, a drought resistant plant, is able to survive.

Dominant	Heather (Calluna vulgaris), replaced by bilberry (Vaccinium myrtillus) on better drained hill tops.
Drier areas	Bell heather (Erica cinerea) Crowberry (Empetrum nigrum)
Wetter areas	Cross leaved heath (Erica tetralix)
Wettest	Sedges (Carex spp) Sundew (Drosera rotundifolia) Cotton grass (Eriophorum angustifolium) Common butterwort (Pinguicula vulgaris) Bog moss (Sphagnum spp)

Areas of mineral soil, without the thick overlaying layer of peat, are indicated by the dominance of wavy hair grass (Deschampsia flexuosa). This, like the plants listed above, is not resistant to trampling, but the soil beneath does at least provide a firm base on which a path can be constructed. If carefully routed and pleasant to walk this should concentrate use and protect the surrounding vegetation. Under very heavy recreational pressure, large areas of wavy hair grass can be destroyed, followed by erosion of the mineral soil, as has occurred on the plateau edge of Kinder Downfall in the Peak District.

Molinia moor

This occurs on waterlogged clay soils in the Pennines, Wales and Scotland.

Dominant   Purple moor grass   (Molinia caerulea)

Wet areas   Cotton grass   (Eriophorum angustifolium)

Mat grass   (Nardus stricta) indicates the areas not subject to waterlogging, but which conversely are not usually suitable for routing paths. This is because they occur either on steep slopes or in damp flushes, where moving water provides sufficient aeration for the roots. Paths are better routed on the areas of purple moor grass, providing surface drainage where possible, and crossing the damp flushes with simple bridges or boardwalks.

Mat grass and purple moor grass have rarely created peat deeper than 300mm, and usually have their deepest roots in the soil horizon. As the peat becomes deeper, and so unsuitable for paths, cotton grass takes over. Better drained areas merge into grassland, dominated by mat grass.

Cotton grass moor

Dominance of cotton grass indicates deep boggy peat. In Western Scotland it may be replaced by deer's hair grass (Scirpus caespitosus). The 'Mosses' of the Peak District are now dominated by cotton grass, the moss that formed the peat having declined due to pollution of the atmosphere. The only possible way to make a path on cotton grass is by a floating construction such as a boardwalk, bundles of heather or brushwood, or 'Terram' overlaid with surfacing material. Costs in time and materials are very high, and constant maintenance is needed if the path is to be kept in good condition.

Drier areas may occur on slopes where water has cut channels through the peat, leaving raised islands which develop vegetation composed of the other moorland species mentioned above. These 'hags' are not suitable for paths because of their configuration, and their great susceptibility to erosion.

GRASSLAND

All grasses are in some degree resistant to trampling, because growth occurs from a point protected by the sheath, and is not on the tip of the shoot. Grasslands are maintained by trampling, grazing and burning, and in their absence develop scrub and tree cover. Thus grasslands are usually suitable for paths.

Summary of heath and moorland species

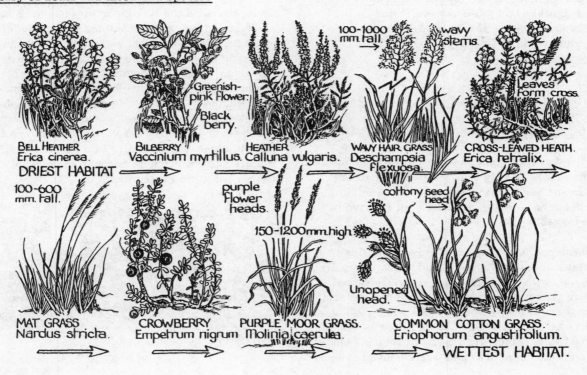

BELL HEATHER
Erica cinerea.
DRIEST HABITAT

100-600 mm. tall.

MAT GRASS
Nardus stricta.

BILBERRY
Vaccinium myrtillus.

Greenish-pink flower.

Black berry.

CROWBERRY
Empetrum nigrum.

HEATHER
Calluna vulgaris.

purple flower heads.

150-1200mm.high.

PURPLE MOOR GRASS.
Molinia caerulea.

100-1000 mm. tall.

wavy stems.

WAVY HAIR GRASS.
Deschampsia flexuosa.

cottony seed head.

Unopened head.

COMMON COTTON GRASS.
Eriophorum angustifolium.

CROSS-LEAVED HEATH.
Erica tetralix.

Leaves form cross.

WETTEST HABITAT.

Neutral grasslands

These occur on the fertile clays and loams of central and southeast England. The presence of the following species may indicate seasonal water-logging.

Wettest areas	Rushes (Juncus spp)
	Sedges (Carex spp)
	Marsh marigold (Caltha palustris)
	Meadowsweet (Filipendula ulmaria)
	Lady's smock (Cardamine pratensis)
	Tufted hair grass (Deschampsia
Drier areas	caespitosa)
	Creeping buttercup (Ranunculus repens)

Note that field woodrush (Luzula campestris) is not indicative of wet soil, and is found abundantly on dry barren pasture land. It flowers during April and May.

Chalk and limestone grassland

These are mostly well drained, but patches of clay can occur on the top of chalk escarpments and between the strata of outcropping limestone. These areas, and damp flushes and springs, will be obvious by the lusher growth, with some of the species listed above.

Upland grassland

These occur on the older rocks of the Pennines, Wales, Scotland and the Lake District. The dominant grass is common bent (Agrostis tenuis), with purple moor-grass (Molinia caerulea) in the wetter areas. Damp flushes are indicated by sedges and rushes. If ungrazed, moorland vegetation tends to take over.

WOODLAND

Conifer woodland

Woodland of Scots pine occurs naturally in Scotland but elsewhere is only in plantations, or where wind borne seed causes spread from plantation onto heathland. Scots pine does not survive in waterlogged soils.

A sparse conifer woodland will have similar characteristics as the heathland which it usually replaces, with paths eroding to a well drained sand or gravel base, and with erosion problems on slopes. In dense woodland all other plants are shaded out, and the woodland floor becomes covered with a thick dry layer of needles. Once sufficient trees or branches are removed to make

a path, there are few problems on a well drained soil, but any tendency to waterlogging is worsened by the exclusion of sun and wind.

A significant problem of routing paths in dense conifer woodland is the fear many people have of getting lost, and which needs careful design and waymarking to overcome (see p33).

Birch woodland

This is a habitat in transition from open heath to oak or beech woodland, and thus is very variable. However, its origins in heathland mean it is usually on well drained sandy soil, with few problems for path management. Damper birch-woods may comprise hairy birch (Betula pubescens) instead of silver birch (Betula pendula), but local drainage patterns are indicated by the presence of damp heath and moorland species.

Ashwoods and beechwoods

These are found on well drained chalk and lime-stone soils, and in some cases ashwood may be a stage in the succession to a beech climax community. Mature beechwoods in particular are ideal for walking, because the dense shade cast by the early emerging leaves, and the thickness of the decaying humus prevents a shrub layer developing, so one can walk about freely. Any patches of boulder clay cause muddy ground in winter, particularly where horse riding is popular.

Dry oakwood

This occurs on the lighter drier soils over sand-stone and slate, and is confined to the south east of England, and the west and north of Britain. The species of oak is the sessile oak (Quercus petraea). Heathland plants such as heather, bell heather and bilberry are found, but bracken (Pteridium aquilinum) is often dominant in the field layer. The soil is dry and shallow, and the shrub layer is poorly developed. Paths are fairly easy to clear and maintain, and erosion is not usually a great problem because the canopy and roots give some protection.

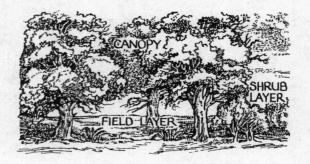

Damp oakwood

This is found on heavy loam and clay soils, mainly in southern, central and eastern England. The dominant tree is the English oak (Quercus robur) but with alder (Alnus glutinosa) in the wettest areas. Damp oakwoods are a very rich habitat with a wide variety of species. Those listed below are typical of the wettest soils:

Marsh marigold (Caltha palustris)
Sedges (Carex spp)
Marsh thistle (Cirsium palustre)
Opposite-leaved golden saxifrage (Chrysosplenium oppositifolium)
Square stemmed willow herb (Epilobium adnatum)
Flag iris (Iris pseudacorus)
Meadowsweet (Filipendula ulmaria)

The main problems with path construction are clearance and drainage. There is little erosion because of the thick soils, lush vegetation and generally flat or gently undulating ground. Although the thick canopy excludes the drying effect of sun and wind, this is partly compensated for by the deep and extensive root system of the English oak, which when in leaf, acts as a perfect sub-surface drainage system, drawing up large quantities of water from the soil and subsoil.

Artifacts

Look around carefully for any features or artifacts which may yield clues as to how a path was constructed or used in the past.

a   Borrow pits. These small pits may have been used for obtaining surfacing material, and could be a useful source for work in the future. They may also indicate a surface that is now hidden by vegetation or humus, and which could be restored by stripping the top.

b   Changes of level between the path and the adjoining field may be the result of sediment accumulating on the path, and covering a hard surface or blocking drains and culverts.

c   The remains of any sort of stonework such as walls, revetment or large paving slabs indicate previous construction.

d   Oddly shaped trees, such as those with thick horizontal sections of trunk near the ground are the result of hedging work done many years previously. Paths often followed these old hedgelines. Remnants of avenues of trees may indicate an old path.

hedgerow bank

Levels of Use

The subject of recreational use of the countryside receives a great deal of academic attention. There have been many surveys both looking at the environmental impact of existing use, and to try and determine the 'carrying capacity' of certain areas. The carrying capacity can be defined as 'the level of use an area can sustain without an unacceptable degree of deterioration of the character and quality of the resource or of the recreation experience'. It is easier to define than to quantify, and its assessment is beyond the scope of this book. However, the principle is useful as a way of looking at the problems and potentialities of a site. It can be sub-divided as follows:

Ecological carrying capacity

This is the amount of use a site can receive without damage to the flora, fauna and soils. For example, a bog has a very low ecological carrying capacity, as very slight trampling will cause damage. In contrast, a well drained grassland will be able to receive fairly heavy use without damage. Quantification of the ecological carrying capacity requires the determination of the original ecological 'base', and of the level of damage which is unacceptable.

Physical carrying capacity

This is the capacity set by physical limits. This could include the width, height and gradient of a path which will affect the type and the amount of use it receives. On many sites, the most important limit is set by the availability of car-parking space.

Perceptual carrying capacity

This is the capacity an area has to absorb use

without appearing crowded to other users. This is related to the physical geography of an area, and to the user's expectations. A remote Scottish mountain may seem spoilt by the presence of another person, whereas an urban park may feel uncomfortably quiet if there is no-one else around. Wooded and hilly landscapes and spaces with convoluted edges (see p34) have a greater perceptual carrying capacity than open treeless expanses. A maximum ecological, physical and perceptual carrying capacity is possibly reached at Hampton Court Maze.

What use is the theory of carrying capacity to the average path worker, toiling to clear a bramble patch or build a flight of steps? Most workers on rights of way on farmland are sensibly concerned only with the practicalities, and not with the fact that they are raising the 'physical carrying capacity'. However, it is a useful discipline when planning or doing work on moorland, coast or mountain, or on reserves of any sort. Many tasks involve repair and restoration, which is a situation where the ecological carrying capacity has already been passed. In most cases one wants to raise this capacity by better routing, drainage or surfacing, without raising the physical carrying capacity.

In this crowded country, perceptual carrying capacity usually needs to be maximised. Careful routing of paths, together with tree planting and provision of one-way circular routes all help to give one a sense of isolation. The 'rule' is to take the limit of an area for recreation as being either the ecological or perceptual carrying capacity, whichever is the lower.

SURVEYING USE

It is best if information can be gathered over an extended period by someone regularly on the site. If this is not possible, try to visit the site on a busy summer weekend, when use is likely to be at its highest. Also talk to local residents and land-owners, who may be able to provide useful information.

The methods outlined below are only intended to give a rough picture of use to plan work for the next year or two on a site. If a long term study is to be done, with comparisons over time, the method must be carefully designed to give statistically useful information. Like the physical survey, it is impossible to give exact procedures as every situation will be different.

This information is based partly on manuscript notes by Neil Bayfield.

Cars

Most fee-paying car parks should have a record of numbers which can be obtained from the manager. The most important thing to note is whether the car-park is filled to capacity at any time, and whether people then turn away, or try to park nearby. By decreasing the amount of parking space, or removing or altering a road sign, use may be lessened. Conversely, the provision of a car-park or lay-by may be vital for the continued use of a right of way. Actual numbers of cars are only useful if comparisons are going to be made with other days or times to year, to see how use changes over time. They do give a good guide to the numbers of people on a path, but such numbers are not particularly useful items of information on their own. Patterns of use are much more important for planning path work.

If any counting is done, use a hand tally, or mark them down in groups of five, as shown below. This is a standard counting system which reduces the chance of error.

Hand tally, or—mark down in groups of five.

|||| |||| |||
ie. total of 13.

Types of users

This is best noted at the car-park or start of the path. Note whether visitors are family groups, serious walkers, dog-walkers and so on. The classification will of course depend on the site. Note particularly any 'special interest' visitors, such as fishermen or rock climbers, and also field parties and youth groups. Such groups can have a great impact on an area, and once this use is established, a pattern tends to be set up and group leaders return again and again. Find out where the groups come from. Also note the numbers of people who only come to sit near their cars or mill about on the initial stages of a path. They might be better catered for with a different sort of facility such as a picnic site.

Routes used

Often this is the most useful part of the survey, particularly in the frequent case where the aim is to channel use along certain paths and restore eroded areas.

1   Find a good vantage point with a clear view of the area.

2 Make a sketch map showing the paths and other features. Long paths may have to be divided into easily recognised sections, for ease of recording. Binoculars are useful for long-distance observation.

3 Count the numbers of people using each path, or sections of path, and whether this is down-hill or uphill use. The counting may need to be split between several observers on busy sites.

4 Note down the details of the survey, including the date, time, weather and exact observation point so that the exercise can be repeated after any path improvements are made. Photos of the view and the observation point will be useful.

The same principle can be used to survey the use of a path at selected points.

1 Select a line at right angles to the path, or use a transect line (see below). Mark the line into sections using stones or small pieces of twig that you can see, but which will not attract the attention of walkers.

SECTION ACROSS PATH

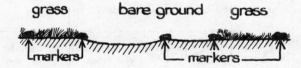

grass   bare ground   grass

markers   markers

2 Record the section used as each walker crosses the line. Without going to extremes, try not to attract the attention of walkers or they will only deviate from the line they are on, or even come over and ask you what you're doing. If you can't hide, at least put your note pad inside a plausible-looking guide book or flora.

3 Note the survey details and location so that the survey can be repeated. If done before and after a task, this is a useful guide to measuring the effectiveness of path improvements.

Difficulties

As you walk the path, note the points where people avoid rough patches, slither down slopes, take short cuts, or appear confused about the direction in which to go. This is very instructive for the observer, as these actions are taken subconscious-ly, and it is such actions that one want to influence. Often observers themselves are too familiar with a site to be aware of these problems,

and seeing them through the eyes of a first-time visitor is the only way of finding them out.

Mingling a little to hear what people say is also helpful. Family groups especially tend to discuss ...."Can we get to the beach this way?" or "Is the waterfall down here or along there?".

INDIRECT RECORDING METHODS

These can be used as a supplement to observed use, and to make detailed records for comparing changes in path width or use over time.

Fixed transect

Select points along the path which are represent-ative of various sections, or where problems are anticipated.

1 Knock in a wooden peg at each end of a line, called the transect, at right angles across the path, and at least one metre outside the existing edge to allow for possible path spread. Depending on the type of vegetation, knock the peg in far enough so it will not attract attention. The pegs are left perm-anently in place. If not already marked, number the pegs with a waterproof pen or by scratching the number in the top.

2 Stretch a tape measure from one peg to the other, holding it taut if the ground is not flat.

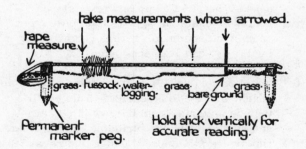

take measurements where arrowed.

tape measure

grass · tussock · water-logging.   grass   grass bare ground

Permanent marker peg.

Hold stick vertically for accurate reading.

3 Note the measurement at any feature you wish to record, such as bare ground, water-logging or untrampled vegetation. Record this in note form or directly onto graph paper.

4 Note exactly how to relocate the transect for future reference.

5 Take photographs both of the actual transects, and to aid relocation.

This exercise can also usefully be done during or at the end of a task, to monitor the effectiveness

of any alterations you make. Unless time is very
short, this type of work also gives volunteers a
welcome change if the work is strenuous.

Walked transect

A more rapid survey technique for longer paths
is to measure the path width (width of trampled
vegetation and width of bare ground) at regular
intervals, for example 20, 50 or 100 paces. This
gives a quick picture of a path along its whole
length. If the path is resurveyed using the same
technique, although different sample points will
be used, the results for a reasonably large
number of samples will be comparable with the
earlier survey.

Other methods

Automatic recording devices and questionnaires
are often used for major schemes. See
Davidson, 1970; Bayfield and Moyes, 1972;
Coker, 1973.

Feasibility

Having come to the conclusion that some sort of
work is needed on the path, note the following:

a   Access. Is it possible to drive materials and
    volunteers to the worksite? If not, how far
    is it to walk? Will there be problems getting
    materials across rough ground, through
    dense vegetation or up and down slopes?
    It may be easier to approach the site from
    another direction, or across neighbouring
    land if you can get permission.

b   Local materials. Is there anything available
    nearby that could be useful? This could
    include material from borrow pits, screes
    and stream beds for surfacing, and boulders
    and stones for steps, pitching and revetments.
    Beach pebbles are useful for path bases, as
    is rubbish and hardcore. Supplies of turves
    may be needed for restoration.

Having collected as much information as you can
on site, and made notes and sketch maps, consider
all the different solutions possible. Take into
account the resources you are likely to have at
your disposal, including numbers of volunteers,
skills, time, machinery and materials. You may
have many months to mull over the problem and
come up with a solution, or the whole job may
have to be surveyed and estimated in an afternoon.
Some people find it useful to write down all the
possible solutions, even ones that seem highly

unlikely, and then successively eliminate them.

Having made your decision, return to the site to
take any measurements necessary for estimating
materials. Decide on the best time of year for
the work. Try to plan out the logistics - where
material should be dumped, the numbers of
people required for each part of the job, and the
number of tools needed.

The sketch below is a copy of one of the many work
cards by Andy Parsons, volunteers' organiser in
the Peak District. These are done in the office
from sketches made in the field, and are for use
on site by the task leader and volunteers. They
are an excellent model to copy. The originals
are done with coloured pencil to make them easy
to follow, and all the instructions and sketches
are kept to a single sheet if possible so they are
easy to handle. The sheet can be protected with
plastic so it will survive rain and muddy land-
rovers.

It is well worth taking some trouble in making the
instructions simple to handle and to follow, instead
of expecting task leaders to cope with sheafs of
complicated notes, often in inclement weather
conditions. Lastly, and most importantly, having
to write and draw out the scheme makes the
designer really think it through.

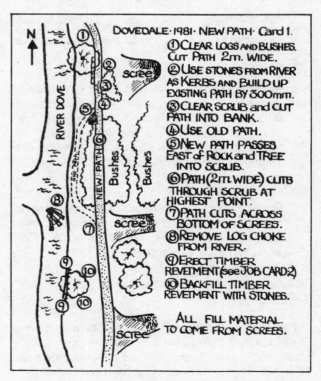

Checklist of Problems and Remedies

This lists the problems most commonly found on paths, with suggested remedies in order of priority. Further information is given on the pages listed.

PROBLEMS		REMEDIES	PAGES
Overgrowth	1	Consider use of machines and herbicides, as well as hand tools, for clearance.	47, 54, 57
	2	Consider removing fences to allow grazing to keep the path clear.	33
	3	Provide waymarking or include the route in a path guide to increase use, which will help keep the path open.	168
Muddy path in woodland or scrub	1	Remove overgrowth and widen path to let sun and wind dry it out.	48
	2	Make camber or cross-fall to shed water into side drains.	75
	3	Build cut-offs, cross drains or culverts as necessary.	65, 68, 69
	4	Lay sub-base, base and surfacing, as necessary.	75
	5	Build barrier if illegal use is causing damage to surface.	164
Soft, deep mud	1	Find source of water and divert it away from the path.	61
	2	Make camber or cross-fall to shed water into side drains.	75
	3	Lay Terram or other sub-base, base and surfacing.	75, 78
Soft mud deposited over firm base	1	Scrape off mud, and dig drains to prevent water running onto path, or the firm base will get covered with mud again.	59
Water flowing across path	1	Divert water at source.	61
	2	Build cross drain or culvert.	68, 69
	3	Build simple bridge.	100
Path crosses patch of boggy ground	1	Divert path.	
	2	Consider ecological value of bog, and either build raised surface, or if necessary, dig side drain.	64
Water running down path	1	Divert water at source.	61
	2	Build cut-offs to divert water off path.	65
Peat, less than 300mm deep	1	Remove peat, use elsewhere for restoration.	19
	2	Fill and surface path as necessary.	76
Wet peat, deeper than 300mm	1	Divert path off deep peat.	
	2	Lay Terram and surfacing material, or lay brushwood.	19, 78
	3	Build boardwalk.	90
Steep slope of bare earth	1	Divert path away from slope.	
	2	Build steps and restore vegetation.	114, 132
Stony, eroding slope	1	Divert path away from slope.	
	2	Build zig-zags, and restore damaged ground.	22, 129, 132
	3	Build steps, and restore damaged ground.	123
Wooden steps with eroded treads	1	Divert path away from slope.	
	2	Re-route the steps.	114
	3	Re-build, surface and drain treads.	119
Braided path (ie split into many parallel tracks)	1	Define and improve best line.	138
	2	Close off other paths, and restore vegetation.	132
	3	Encourage use of defined path.	37
Multiplicity of paths with serious erosion, needing major repair scheme	1	Define and improve major routeways, and close off and restore others.	129
	2	Restore vegetation by seeding, turfing and planting.	132
	3	Provide publicity and information to help get the scheme accepted.	37
Muddy gateways and stile approaches	1	Consider whether barrier is necessary, and if not, remove it.	
	2	Surface, eg with stone pitching or stone causeway.	84, 86

4 Path Planning: location and use

This chapter deals with a variety of topics concerned with the appearance of paths, and the way in which people use them.

Appearance in the Landscape

Surfacing

Most unsealed surfacing materials weather with time until the surface is hardly noticed as being imported. Local supplies from borrow pits and stream beds not only have a better chance of blending with the surroundings, but have a more variable colour and texture than commercial supplies. It is the uniformity of a newly made surface that strikes the eye.

Sealed surfaces such as tarmac and concrete have a uniform texture and colour which makes them obtrusive in the countryside. The appearance can be improved slightly by rolling chippings into the surface of tarmac and roughening concrete before it sets, but both materials should be avoided if at possible. The colour of concrete is slightly more acceptable than tarmac, but it is an awkward material to deal with if it starts to break up. The concrete sandbag path shown on page 83 is useful in the type of situation quoted, and the construction method effectively disguises the fact that it is concrete.

Edging

Avoid using an edging to contain surfacing material, as it will show as a hard line, noticeable even after vegetation partly covers it because of the inevitable straightness of the path. These edgings, such as railway sleepers, are only necessary where the path has be raised, for example above a high water table. In all other cases, it should be possible either to simply roll the surfacing into place, or excavate and fill (see p76) to form the path. Never place an edging merely to define the edge of a path in an attempt to keep walkers on it, as such edgings are not only ugly, but ineffective.

Artifacts

The workmanship seen in the surfaces, walls, stiles and bridges of many old paths is part of their attraction. Most involve time consuming methods, using local materials. Fortunately, this is the type of work most suited to voluntary effort. Modern materials have uses, particularly in ground restoration, but they are then usually covered by soil and vegetation.

Use of traditional materials, however, is not a guarantee of good design, and anything that resembles the 'gardenesque' is inappropriate in the countryside. Artifacts must be functional, simple and robust, with no embellishments. Any that are not functional are unnecessary, anything complicated is inappropriate, and anything not robust enough to stand weather and use is a waste of effort. The following illustrations are some examples of the 'gardenesque' seen on rural paths.

Inappropriate embellishment to wall.↓

edging!

crazy-paving seal.

"Rustic" seat.

← edging unnecessary!

General Principles of Path Design

Quality

Work on paths not only must look good, but it must be well built to withstand time, use and weather. A little amount of sound quality is of far more value than rushing to do a large amount of work of poor quality. Re-routing, surfacing, drainage or steps badly done will all cause more trouble than they cure.

Desire lines and short cuts

Most people, when walking from A to B, will take the most direct route possible. This is called the desire line. The whole system of rights of way is built up of countless desire lines. Problems arise when a path does not follow the desire line, either for legal reasons, or because the land is not physically capable of supporting a path.

Enclosed paths

Narrow paths enclosed by hedges or fences are often a problem as concentrated use rapidly destroys the surface. In urban areas these alleys or 'twitterns' are usually paved or surfaced with tarmac. In rural areas, if the choice exists, it is preferable to avoid fencing in a path. Use is not then concentrated on a single line, and the problem of cutting growth along fence lines is avoided.

The most satisfactory situation is where the path runs through pasture, and is managed as part of the field and kept short by grazing. Once the path is fenced it will either become overgrown, or muddy, depending on the amount of use. Trespass is likely to follow as walkers climb the fence in order to walk across the grazed part of the field. If there is a problem with walkers wandering off the path, it is better to try first by use of signs or marker posts to keep them on the path. If fencing is necessary to separate grazing animals, use the minimum number of strands which will still be stockproof, or fence on one side only, so that as much of the path as possible is still grazed.

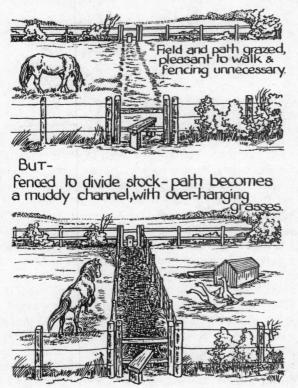

Field and path grazed, – pleasant to walk & fencing unnecessary.

But – fenced to divide stock – path becomes a muddy channel, with over-hanging grasses.

Targets

All sorts of features and artifacts can act as a 'target' on a path. A target is anything that takes the walker's interest, and may be a natural feature that appears in view along the path, or something seen on a map or heard about from friends. The most frequent targets are high land, views, water and ancient buildings, but rare plants or birds, famous and infamous residences, ley lines and sighting places for UFOs can all alter, often unexpectedly, the way in which a particular path is used. Major problems arise when the target is off the path, and trespass becomes a recurring problem. Most targets are immoveable, but at least by identifying them one is able to understand why and how a path is used, and can then take action to re-route the path or screen the target from view.

Targets more easily dealt with are the artifacts that may occur along or near a path. These include information boards, seats, litter bins, cairns, collection boxes and even piles of rubbish which may attract the walker's eye and magpie instincts. If carefully placed, they can be not only functional, but useful in drawing people in certain directions. More often though, they are just a focus for trampling and erosion.

The following stone, placed by an eccentric gentleman in the 1880s, still exists at Durlston Country Park, Swanage, Dorset. The irony is further increased by the fact that the lettering is now so worn it is necessary to get very close to read it.

Claustrophobia and agoraphobia

Paths that lead into dark woodland can be very intimidating to some people. This is especially a problem with the conifer plantations that are increasingly being opened to the public by the Forestry Commission. Grassy rides are the most pleasant type of forest walk, being wide and light and having the interest of the forest edge habitat. They often have straight sections and views of features which act as a reference point for walkers (see 'straights and curves' below).

Paths that have to go under the canopy are a different proposition, especially if there is 'no light at the end of the tunnel'. A path clearly defined by surfacing, such as the woodchip path described on page 81 is the most satisfactory method of encouraging use, so that people are not worried about losing their way. Other methods include various types of waymarking, and having copies of maps available which can be taken on the walk. To those used to the countryside and the forest this may appear a minor problem, but it is in fact enough to discourage many people who are at all wary. The provision of targets (see above) in otherwise rather monotonous forest is useful both to relieve the eye and encourage the feet, and to help recognise the way back.

In the example below, it was found, after the car-park had been constructed, that walkers were reluctant to set off into the dark woodland. Extra waymarking was therefore provided.

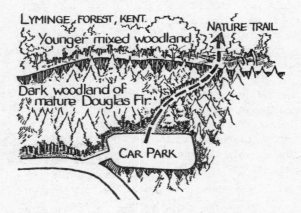

At the other extreme, some people are hesitant about setting off across wide open spaces if there is no target at which to aim. The edges of a space or where two habitats meet will always have more interest and appeal than the middle, and unless a desire line exists across it, most people will tend to keep to the edges.

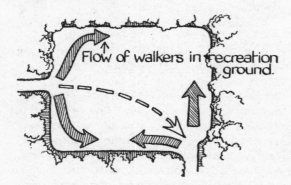

The understanding of how people use space is helpful in the management of recreation areas. Open spaces in parks and picnic sites are usually designed with convoluted edges to give the maximum edge where people will walk or sit. This links with the principle of carrying capacity (see p27). Varied landscapes with trees, water and plenty of 'edge' will have a higher perceptual

carrying capacity, and probably ecological carrying capacity. than monotonous landscapes.

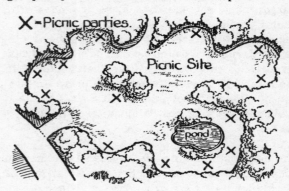

Straights and curves

The great parks of the 16th and 17th Centuries were laid out in the grand manner with long straight rides and avenues, which was particularly suited to the flat landscapes of France, where the designs originated. A pattern frequently repeated was that of the patte d'oie or goose foot, with three, five or seven paths radiating from a semi-circle.

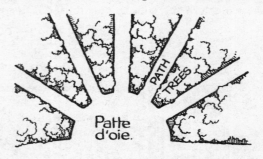

These patterns can still be seen in places such as the Forest of Fontainebleau. They were copied in England, but never very successfully, because the impact of straight lines was lost in the rolling English landscape. A quite different style emerged in Britain, in the work of Repton, Capability Brown and others, which followed the maxim that "Nature abhors a straight line".

Most formal and informal parks contain some of each style. The view along a straight avenue gives a feeling of grandeur and space, and the winding paths give seclusion and interest. Most people tend to walk a short way up an avenue, and then turn to explore the more intimate paths on either side. The same pattern can be seen amongst visitors to a cathedral: not many venture straight up the nave, but walk up the aisles to either side, and across the transept to the chancel.

A combination of straight and curving paths works well in forest, particularly if the land is flat or the planting monotonous. The straight avenues give a sense of space and reference point against getting lost, and curving paths give interesting walking. For forest management, straight avenues give visual as well as physical access, with towers for fire watching and deer culling often positioned at junctions of avenues. A division of recreational users can also be made; the curving paths under the canopy reserved for walkers only, and horseriding allowed on the avenues or forest roads.

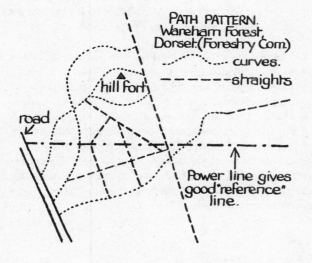

PATH PATTERN.
Wareham Forest,
Dorset.(Forestry Com.)
---------- curves.
— — — straights

hill fort

road

Power line gives good "reference" line.

Dimensions

Visualising dimensions of paths and steps outside can be difficult, and should not be judged from interior dimensions. In the same way that the foundations of a house look ridiculously small before the walls are built, the widths of interior stairs and corridors will look too small if transferred to the great outdoors. Wide paths and steps not only function better, but 'sit' more easily in the landscape.

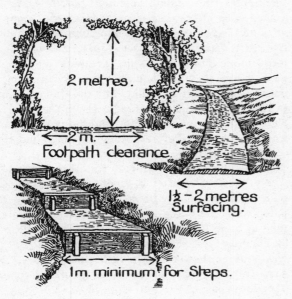

2 metres.

← 2 m. →
Footpath clearance.

1½–2 metres Surfacing.

1m. minimum for Steps.

Permanent or temporary

Construction work can sometimes be avoided by using paths in rotation. This system is used in gardens and formal parks, where grass paths are closed off as soon as signs of wear show. The grass is then either left to recover naturally, or hastened by the use of fertilisers, seed or spiking. A similar method can be used on grassland or downland, especially on hill forts and other ancient monuments, where any permanent construction work is particularly undesirable. At Uffington White Horse, Oxfordshire, simple barriers of wooden posts and single strand wire are used to control the flows of visitors and close off areas becoming worn. These barriers do the job without marring the appearance of the hill fort, and are easily moved as necessary. Here the aim is to allow access without allowing the formation of paths, by encouraging an even use over the whole area.

Temporary paths can also be used in meadow grassland, by cutting a path through the tall sward, and mowing it as necessary through the summer. If this gets worn, a different route can be chosen the next year. This system is also good for protecting meadow plants, as only a few people will be tempted off the mown path, and any rare plants will be much less conspicuous than if surrounded by a protective fence, which will only attract attention. These type of paths are cut each year at Upton Country Park, Poole, Dorset, to make a network of winding paths in meadow and parkland. It also raises the perceptual carrying capacity (see p27) by giving a sense of seclusion. The same grassland mown short, although with a much greater area available for use, would probably hold fewer people as the sense of seclusion would be lost.

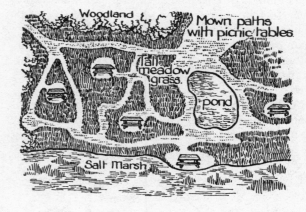

Woodland   Mown paths with picnic tables

Tall meadow grass

pond

Salt Marsh

Circular paths

Circular paths are usually popular as no retracing of steps is necessary. They are also easier to manage than a network of paths, as peoples' behaviour is more predictable, especially if the route is one way only. Often circular routes are naturally walked in one direction only, if for example a target is in view from the car park. One way use can be encouraged by screening or placing of targets, by the relative placing of the 'entrance' and 'exit', and by notices and gates. One way use has the advantage that the path can be narrower, and waymarking and provision of leaflets is simplified, as they need only indicate or describe the route from one direction.

Another interesting point about circular paths is that length of stay for any visitor is then fairly predictable, and this can be linked to car park provision. Especially if there is no view from the car park, the 'turn-around' of cars will be the time it takes to walk around the circuit.

Short sections of 'dual carriageway' are useful for reducing trampling and erosion.

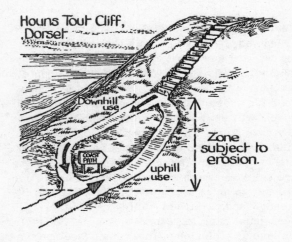

Side paths

One method of decreasing the number who visit a fragile habitat or feature is to alter the line of the path so that it is necessary to walk along a side path or cul-de-sac to reach it. Depending on the situation, some people will then fail to notice the 'target', and others will not bother to go out of their way. Screening with planting may be effective, although a partly concealed path can sometimes have more attraction than one which can be seen along all its length.

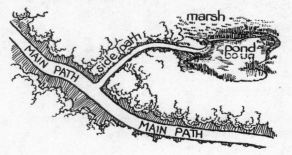

Re-routing

One of the most difficult jobs to do successfully is to re-route a path, where the original route has become damaged or eroded due to over-use. It is much better to 'get it right first time', than to have to try and change existing patterns of use. In nearly every case the existing use will follow the desire line, and physical barriers will be needed to block it off. Deep, wet mud is about the only physical condition on a path which discourages walkers from following the desire line.

If it is not possible to shut off the old route, there is no point in creating a new one. There is also no point in permanently re-routing a path if the same problem is going to occur as the new path is used. This just causes a gradual degradation of the area, as each new 'permanent' route is abandoned in favour of another. If it is not possible to re-route a path along physically more resilient land, then if access is to be maintained, the only alternative is to do some sort of constructional work.

Upland Paths

Uphill and downhill movement

There is a physical and a psychological difference between the uphill and downhill use of slopes, both of which tend to make downhill use the more damaging. As this damage increases with increasing slope, it is better on a one-way path to try and route it up the steepest slopes, and down the less steep slopes.

Physically, the added impetus of downhill use makes each step cause more damage than does the slow and careful placing of each step when walking uphill. The foot tends to slide, and to dislodge loose stones or earth. Walkers tend to move more carefully and slowly when climbing uphill, because it is tiring, and the line of sight is limited to the area immediately in front. Short cuts are less obvious, and the plodding walker is more likely to follow zig-zags and steps.

Downhill use is much more difficult to control. The view down is often clearer, and the line of path can more easily be seen, along with any potential short cuts. People going downhill also tend to be in a hurry, either racing each other, or running for the fun of it, or hurrying to get down before the light fails or the pub closes. Not surprisingly, the majority of hill-walking accidents happen as people are going downhill.

As well as the vertical force of the toe and heel, the shearing force of turning a corner is significant, as people turn sharply to zig-zag up or down a slope. The maximum vertical force recorded in straight walking is about 120% of the body weight, but rises to 150% in turning a corner (Huxley, 1970, p4).

Controversy

Work is needed on many mountain paths in order to make them of a standard which will prevent existing or increased levels of use from damaging slopes. Such work is the subject of some controversy, centred around the three questions outlined below.

1   Will improving the path cause the rate of use to increase?

2   Will improving the path change the type of use, and in particular tempt the ill-equipped or inexperienced walker to venture further up the mountain than he can safely cope with?

3   Will the improvements mar the appearance of the hillside, and the experience of being in a 'wild' landscape?

To answer these three questions:

Improving the path can eventually increase the number of people who use it. However, the presence or absence of road signs, car parks, information centres, and the extent of publicity are likely to be much more important factors than the physical state of the path. The situation cannot be made worse, as the path may already be in a state where damage is rapidly spreading.

Will improvements tempt the ill-equipped? Possibly, but it is the case that hundreds of ill-equipped people already climb mountains, as shown by the numbers who reach the summits of Ben Nevis or Helvellyn in heels or flip-flops. This is the opportunity to make positive improvements: by proper management of the path so that it can cater for the number wishing to climb it, and by education and information to encourage the public to be properly prepared and to treat the mountains with respect. This can be combined with the provision of better low-level paths and facilities, to cater for those who know no other than the famous and eroding mountain paths.

The actions of organised groups have a great impact, as they not only converge on the familiar peaks, but reinforce the popularity of these paths as group members return on family visits, or even return years later when they themselves are group leaders. To explore lesser known, if less spectacular paths would be more of a challenge, and less environmentally damaging.

There are several points relevant to the question of the marring of a wild landscape:

a   An eroded path is already a scar in a wild landscape.

b   The designs in this handbook recommend the use of natural materials and labour intensive methods which blend with the landscape.

c   The landscape of Britain is not a wild one, it is semi-natural, and similar types of work done in previous centuries are now an admired part of it. There is no reason why 20th Century improvements should not likewise mellow with time.

d   The days when one could walk alone through the now popular peaks of upland Britain are probably gone. The people who lament those days perhaps also lament all the other changes brought about by car-ownership, motorways and five day working weeks. It is worth remembering Nan Fairbrother's description in 'New Lives, New Landscapes' of the Lake District just before World War II, which is one of thanks, not lament, "....how lucky we were to coincide with the tiny span of social history when such conditions existed: a generation before we should never have reached the hills as playground, a generation on and everyone will be there." And so they are.

Keeping Users on the Path

In most situations, walkers only have the right to walk along the line of the right of way, which should coincide with the path. There are therefore legal reasons for encouraging walkers to keep to paths, and often there are agricultural and environmental reasons also.

In some areas walkers have the legal right to walk anywhere, and if the land is suitable and no damage will occur, there is no reason to keep to the path, if indeed one is necessary at all. Where paths are necessary to protect the land, it is also necessary to keep people to them.

The following section suggests various methods of obliging or encouraging walkers to keep to the path.

OBLIGATORY METHODS

These should only be necessary where it is essential for legal or safety reasons to keep people on the path, or where trampling must be excluded from recently restored areas.

Fences and barriers

Fences along the sides of paths are not recommended as they are expensive to install and greatly increase the maintenance requirement of the path (see p33). They also need to be unclimbable if they are to be totally successful, and are then likely to be very unattractive. Fences may be necessary for safety reasons.

Barriers can be used across paths and entrances either to keep out certain users, or to seal off paths where a new route has been made. They may also be needed to block off short cuts. To be effective they have to be unclimbable if the aim is to close the path completely. Chestnut paling fence or un-tensioned chainlink fencing is suitable for closing narrow gaps as a temporary measure unti planting obscures the path. Dead plant material is useful as it can form an impenetrable mass. Use thorn or gorse if possible, and lay with the butt ends away from the path so the walker is faced with the mass of twiggy growth.

Signs

'Trespassers will be prosecuted' is meaningless as trespassers can only be prosecuted for causing damage. 'Private' or 'Keep Out' are too familiar to be effective. If signs have to be used, an advisory message is usually better.

ADVISORY METHODS

On open land it is impractical to use fences or barriers to oblige people to keep to paths, and instead methods must be used to win the walker's respect for the reasons for keeping to the paths.

Waymarking

This is the standard method of advising walkers of the route of the path, and is especially important on farmland. Waymarking in upland areas is more controversial (see p175).

Signs

Messages which explain, eg 'Erosion control: please keep off' and 'Stock in field: please keep dogs on leads' are preferable to ones which merely prohibit. As shown at Kynance Cove, information displays explaining restoration and path work are very helpful in getting a scheme accepted, and are appreciated by the public (O'Connor, Goldsmith and Macrae, 1979).

Occasionally the use of sharp messages such as 'Wrong way: go back' are useful on short cuts or closed-off paths.

Educational

The Country Code includes the advice 'Keep to public paths across farmland' and 'Use gates and stiles to cross fences, hedges and walls'. The Lake District and other national parks produce literature aimed primarily at school and youth groups, which includes advice on walking mountain paths and the importance of keeping to zig-zags. This will hopefully be extended so that all visitors to the uplands are made aware of this advice.

OTHER METHODS

Planting can function in several ways to direct the movement of walkers. Suitable plants with their growing requirements are listed on page 145.

Screens

Planting of thick, preferably evergreen bushes or trees can be used to screen a target which is off the path, or to screen potential short cuts. The latter will need to be combined with some sort of fencing to allow the plants time to establish.

Barriers

Hedges of thick, usually thorny vegetation are the traditional method of making barriers. Stock and people-proof hedges need a lot of maintenance to be kept in good condition, but informal planting of thorn, holly, wild rose, bramble or gorse can be used for a fairly maintenance free barrier which will discourage people. They are best used for blocking short cuts, or planted in a mass to discourage access across an area. If used alongside a path, bramble and wild rose will need cutting back each year to keep the path open.

Diversionary planting

An interesting method was used at Cwm Idwal, North Wales, to keep walkers to a broad curving path and discourage them from short cutting. Clumps of rushes were planted and succeeded in keeping users to the path where fences and other barriers had failed. The explanation is that rushes immediately suggested wet ground, which people instinctively avoided.

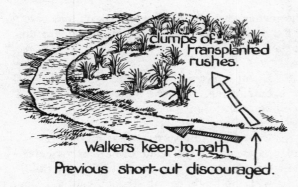

Walkers keep to path.
Previous short-cut discouraged.

Boulders, hollows and targets

Boulders are effective at keeping walkers on the path so long as they are carefully placed and landscaped to look entirely natural. This has been clearly demonstrated by work done on the nature trail path at Ben Lawers, Tayside, where boulders have been sited intermittently along the edge of the path, or at the start of short cuts. They are carefully placed with the weathered and lichen-encrusted side upwards, set into the ground at least 100mm, and the turves replaced around.

These boulders appear to work because they are completely unobtrusive, and without being aware of it, the walker follows the intended path, even though the boulders are low and easily stepped over. On many similar sites, barriers have been made of stone roughly piled up, like small walls. These look unnatural, attract attention, and without probably meaning to be perverse, the walker steps over and continues on the 'closed-off' path. The 'trick' appears to be to not allow doubt as to the path's route register in the walker's mind.

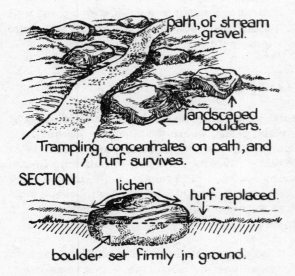

path of stream gravel.

landscaped boulders.

Trampling concentrates on path, and turf survives.

SECTION

lichen

turf replaced.

boulder set firmly in ground.

In many situations a depression or a sharp drop from the edge of a path is more effective than a fence or barrier. It has been shown that if faced

with a depression in the ground ahead, people are more likely to walk around than if there is a mound of similar volume.

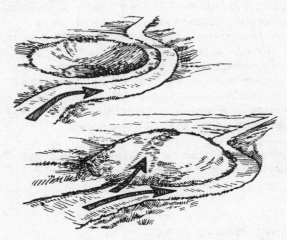

This rather theoretical point can be adapted to discourage short-cutting on zig-zags, so the downhill walker is presented with a hopefully intimidating drop. A deep drainage ditch, even if usually empty, discourages walkers from stepping off the path.

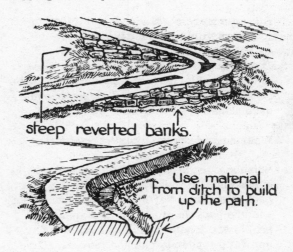

steep revetted banks.

Use material from ditch to build up the path.

Targets, if skilfully placed, can draw the eye and attract the walker along a certain route. Views are the most reliable target, but an information board or seat may also serve this purpose.

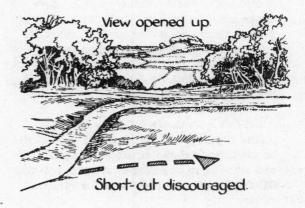

View opened up.

Short-cut discouraged.

5 Safety, Equipment and Organisation

The following information is basic to many types of voluntary footpath work. The main problems arise from the linear nature of the work site, and the fact that it is usually inaccessible to vehicles. The leader must be aware that it will not be possible to supervise all the work all the time if the group is spread along a length of path, and should explain details accordingly. It may be necessary to duplicate first aid equipment and tools. An inaccessible site will require the carrying in of tools and materials, and may lead to difficulty in the case of an accident.

Further information on the organisation of specific jobs is given in the relevant chapter.

Safety Precautions

GENERAL

a Have a first aid kit to hand. A suitable kit is listed below.

b All volunteers should have had a tetanus injection within the last five years.

c Do not work in soaking rain. Once gloves, tools and materials are sodden, the danger of accident increases, and slopes become hazardous.

d Wear suitable tough clothing (see below).

e When moving heavy weights, bend your knees and use your leg muscles, not your back muscles. Never try to lift more than you are capable of, and if two or more people are together moving something, appoint a leader to call the moves. Time spent on planning the stages of moving a heavy object is not time wasted.

f In most cases you will be working on a public path, and the safety of other path users must be kept in mind. Stop work to allow walkers and riders safe passage if you are using potentially dangerous tools. Warning signs may be helpful, and can incorporate publicity for the group doing the work.

g All BTCV volunteers are covered for public liability and personal accident while on task, and groups affiliated to the BTCV can arrange to be covered by their policies. Other groups should arrange for similar cover.

TOOL USE

The leader should give a brief talk on tool safety at the base or van, before the group walk to the worksite.

a Never use a tool with which you are unfamiliar. Ask the leader to explain and demonstrate its use.

b Carry tools by your side at their point of balance with the head of the tool forward, where you can see it. Hold edged tools with the blade down. Do not walk too close to others, or try to overtake on narrow paths. There may be an exception if crowbars have to be carried a long way up a hillside, when carrying across the shoulder is less tiring. Keep well away from others.

c Keep a safe working distance from others.

d Do not leave tools lying around where they may be stepped on or lost. At each work site have a base for the tools away from the immediate working area, and out of the way of walkers. Beware of leaving crowbars on a slope where they may slide or roll down.

e Tools are safest to use when sharp and well maintained (see p45). Never use tools with loose or damaged handles.

UPLAND PATHS

Special precautions should be taken when working on upland paths, where the mountain weather, rough terrain and inaccessibility combine to make potentially dangerous working conditions. Leaders must be aware of the symptoms of exposure, and be able to give first aid. They should check local rescue arrangements in case of accident or exposure.

a Take spare clothing. The weather can change rapidly in the mountains, and a sunny day may suddenly become misty and cold.

b Include survival items in the first aid kit as

an injured person will rapidly lose body temperature. Include an exposure bag and high glucose food such as dextrosol or Kendal mint cake.

c   Unless the weather is very warm and settled, it may be better to plan on having a short lunchbreak and finish earlier in the afternoon than is usual on a lowland task. Work on upland paths is usually arduous, and the walk up combined with exposure to wind or sun may tire volunteers more than they expect. Accidents are more likely to happen when people get tired and lose concentration. The length of the walk back to the vehicle must also be kept in mind.

d   Take great care on the way down, and do not rush or race back. More accidents happen in hill-walking on the way down than on the way up.

e   The moving of boulders must be done with great care both to volunteers and walkers, especially if it is necessary to take boulders from above the line of the path.

Clothing

a   Wear overalls, a boiler suit or close-fitting clothes. Loose clothing and scarves are dangerous when working with tools or moving heavy objects.

b   Heavy leather work boots with vibram or nailed soles and steel toe-caps give good grip and protect the feet. If wellingtons need to be worn, preferably wear the type with steel toe-caps.

c   Gloves with a gauntlet are very useful when clearing brambles or nettles, but do not wear very stiff gloves on the hand holding an edged tool as it makes the handle difficult to grip. Leather gloves are also useful protection when handling heavy boulders, but must fit well and not slip. Wear heavy duty rubber gloves when handling creosoted wood. Canvas or leather gloves will only absorb the creosote and retain a permanent aroma that transfers to the hands!

d   Wool trousers or breeches are useful in wet or cold weather, particularly in the uplands, as they retain warmth even when wet.

Tools and Accessories

FOR ALL TASKS

First aid kit. Have one available at all times. Pinched or crushed fingers are likely to be the most common problem, along with cuts and thorns if clearance work is being done. Include at least the following:

    Tweezers
    Packet of needles
    Matches
    Large plain wound dressings (BPC No. 15)
    Box of medium size porous plasters
    Gauze dressings
    Cotton wool
    25mm cotton bandage
    100mm crepe bandage
    Triangular bandage
    Scissors
    Safety pins
    Eye lotion and eye bath
    Insect repellant
    Mild antiseptic cream
    Antihistamine cream for insect bites
    Sunscreen cream
    First aid book

Many of the tools listed below are included in the BTCV tool catalogue, which is issued to all BTCV groups. Names and addresses are given in Appendix B, where indicated, for brands that are particularly recommended, or for specialist tools. Groups buying other than through the BTCV will find the best selection in the ranges of agricultural and contractors tools (p181 5.1, 5.2). Metal YD handles with wood grips on forks and spades are recommended for durability.

Tools are listed under different types of work, but many tasks will require a selection from each list.

CLEARANCE

a   Slasher. For use on rank vegetation such as brambles, nettles and light scrub.

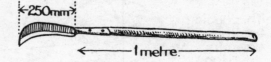

b   Grass hook. These are obtainable with a cranked handle to keep the user's hand clear of the ground, but can then only be used as designed for either right or left hand. Flat hooks can be used in either hand, or for 'back-handed' use. A crooked stick helps to position the vegetation for cutting.

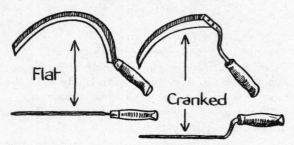

c   Scythe. These are obtainable with either a long blade for grass cutting, or a short blade for brambles.

d   Swipe. This is marketed as the Eversharp scythette, and is also known as a 'turk' or 'swizzle stick' (p181 5.3). It is swung like a golf club to cut grass, nettles and other non-woody growth. The blade can be replaced if broken.

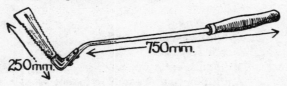

e   Secateurs. These have either a scissor or anvil action. Ratchet models are available which can cut branches up to 20mm diameter (p181 5.4).

f   Toggle loppers and pruners. Toggle loppers can cut branches up to 30mm diameter, but may bend out of alignment if used beyond their capacity. A small right angled metal bar welded to the blade helps strengthen it. Ratchet pruners can cut branches up to 50mm diameter (p181 5.4). Short handled pruners are useful for fitting into a rucksack (p181 5.6).

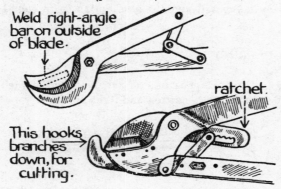

g   Bow saw. The triangular saw is useful on small branches or where space is confined. The D shaped handle gives greater clearance for the blade, and is needed for cutting large timbers.

h   Pruning saw. This is used on close-branched trees where a bow saw will not fit. Handles can be attached to give extra reach. Folding saws which fit in a rucksack are obtainable (p181 5.7).

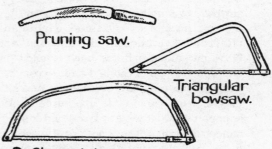

i   Cross-cut saw. Although this tool has been little used since the advent of the chain-saw, it is appropriate for clearing fallen trees on inaccessible paths, where effort would be wasted on carrying in a chain saw and fuel.

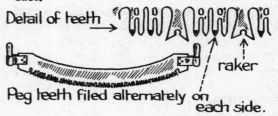

j   Axe. Useful for clearing fallen trees and preparing local timber for use in steps or revetments. Use a heavy (2025gm) axe for felling and a light (1140gm) axe for snedding or trimming. Axes require practice and skill to be used effectively.

k   Wedges and sledgehammer.

l   Mattock. Grub ended for cutting through roots.

m   Single-edged billhook. For light scrub clearance, snedding and brashing.

n   Sharpening stones (see p45).

o   Fire-making kit. Kindling, newspapers, matches, pitchforks, sump oil or tyre.

DRAINAGE AND SURFACING

a   Heavy garden spade, treaded.

b   Trenching spade with triangular, slightly bevelled blade. A versatile tool, as it can be used for both digging and shovelling.

c   Heavy digging fork, with YD handle.

d   Square or taper-mouth shovel.

e   Steel garden rake.

f   Pickaxe.

g   Pick ended mattock for general use, and grubbing mattock for roots.

h   Crowbars, length 1200mm or 1500mm.

i   Heavy duty builders wheelbarrow, with pneumatic tyre.

j   Buckets. Heavy duty rubber or canvas, but not plastic or metal which break easily. Old buckets with holes are useful for collecting stream material for surfacing.

k   Screens, for grading surfacing material.

l   Drainage rods and head attachments for clearing blocked drains and culverts.

m   Punner, for tamping down surfacing.

Specialist ditching tools

a   Rutter, for use in deep peat and stone-free mineral soil.

b   Hack, for pulling cut turves out of ditches.

c   Draining spade, with long handle and blade up to 450mm long, for cutting deep trenches.

d   Long handled spade with no 'lift', for cleaning sides of trench.

e   Tile scoops, for cleaning loose earth from the trench bottom.

CONSTRUCTION WORK

This includes the construction of stiles, steps, bridges and boardwalks.

a   Carpentry tools. Panel saw, claw hammer, lump hammer, adjustable spanner, spirit level, screwdrivers, brace and bits, chisels, surform, wrecking bar, steel tape, mallet, junior hacksaw, wire cutters, pencils.

b   Square ended shovel for mixing concrete or mortar.

c   Mixing board and bucket. A sheet of old lino makes a useful portable mixing board for inaccessible sites. Alternatively, concrete can be mixed by adding water to the dry materials in a plastic sack and rolling it over and over.

d   Post hole borer. Bucket type for sands and gravels, and auger type for clays and loams (p181 5.2).

e   Spade with a curved blade for hole digging. A draining spade is suitable (p181 5.2).

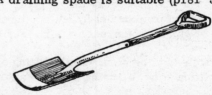

f   Shuv-holer for the removal of debris from large post holes. A modified trowel is useful for removing debris from small post holes.

g   Scoops, of cut-down plastic squash bottles, for scooping water from post holes in wet conditions.

h   Mell, maul or Drive-all for driving in wooden posts.

i   Sledgehammer, for driving in metal posts.

j   Fencing tools. Bolt croppers, fencing pliers, strainer.

k   Punner, for ramming fill around posts. For narrow post holes, a useful tool can be made out of 50mm iron pipe with an elbow section at the bottom.

l   Cold chisel and bolster.

m   Winch. The Tirfor TU16 has a 750kg safe working load, with 18m of 11.3mm diameter galvanised maxiflex cable with a 1620kg safe working load.

n   Walling or brick hammer, useful for building stone revetments.

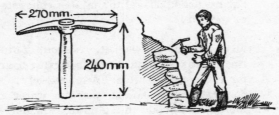

o   Fertiliser bags, slit open, for keeping backfill tidy while hole digging.

MISCELLANEOUS

a   Safety helmets (BS 5240:1975). For any task involving tree felling, lopping, transport of material by helicopter or aerial ropeway, or construction work in gorges.

b   Goggles, to be worn when splitting or drilling rock, or using a brush cutter.

c   Waymarking tools. See page 173.

d   Jerrican carriers. By cutting away the corners, these can be adapted to carry small tools and materials.

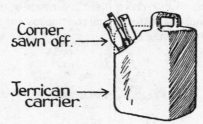

A mobile kit

Although it is generally considered permissible to carry light clearance tools, such as secateurs, to cut back surplus growth encountered whilst walking a right of way, anything beyond this must be done with the prior agreement of the land-owner. The list below is a selection of tools and materials that fit in a rucksack, and which are

useful for an individual or small group who have permission to maintain a length of path.

a   Secateurs.

b   Grass hook.

c   Folding saw.

d   Gauntlet gloves.

e   Baler twine, wire, hammer, nails.

f   Rag or stiff brush for cleaning waymarks.

Machinery

Decisions over use of machinery are usually dictated by factors such as cost and access. Footpath work can involve hard labour, especially in transport of materials, which machines can usually do more efficiently. When considering the use of a machine on a task involving volunteers, bear in mind the following points.

a   The machine must only be operated by a skilled or qualified person.

b   Will the machine present a hazard to volunteers? Can the machine work be finished before the volunteers arrive on site, or if not, can the volunteers work at a safe and quiet distance from the machine?

c   If the job is only possible with the use of a machine, is it really suitable for volunteer labour?

d   Will the volunteers simply be 'slaves' to serve the machine, or will it take away much of the drudgery and leave volunteers free to do more skilled and constructive work?

e   The conservation of the landscape is a primary aim. Will the machine damage more than it helps mend?

Details of suppliers are given where indicated in Appendix B (p181). Most machines are also available on hire. Look under 'Contractors Plant and Machinery Hire' in the Yellow Pages telephone directory.

a   Chain saw. Use of the chain saw for felling and snedding is described in 'Woodlands' (BTCV, 1980).

b   Brush cutter. See page 55.

c Knapsack mower. See page 55.

d Flail mower. See page 55.

e Power earth auger, for making post holes (p181 5.8).

f Rock drill, such as the Pionjar 120 which is petrol driven and weighs 26kg. It can be carried to inaccessible sites on a pack frame (p181 5.11).

g Dumper truck. On suitable ground that is not too soft, these can be invaluable for transporting surfacing materials. They do not handle as expected by a car driver, and should only be driven by trained volunteers!

h Vibrator roller, for compacting surfacing material.

i Grader. This is a tractor-mounted blade for making cross falls on path surfaces. See page 75.

j Excavator, for clearance, drainage and surfacing work.

k Helicopters. These have been used to move heavy items, such as bridge beams, into remote sites. They have also been used for moving surfacing material, but for safety reasons this should not be done while volunteers are on site, as material can easily fall from the hopper. Charter firms are listed under 'Air charter and rental' in the Yellow Pages telephone directory.

l Argocat. This is an 'All Terrain Vehicle' that has a very low ground pressure, and so does minimal damage to soils and vegetation. The smallest model can carry 285kg (p181 5.12).

m Generator to supply any power tools needed for bridge building. It will normally be hired with tools to suit, but check that the voltage supply is correct for the tools required.

n Compressor. This is needed for heavy duty air-powered tools such as rock breakers. Vehicle access to the site is necessary, as compressors are very heavy.

o Chain saw drill attachment, for drilling holes in wood, steel or masonry (p181 5.10).

Tool and Equipment Maintenance

STORAGE AND GENERAL CARE

a Keep tools clean and dry. Carry a rag with you in wet weather to wipe them off, particularly the handles, which become slippery if wet or muddy. Clean tools after use by scraping mud off and washing blades.

b Oil all metal parts before storing to prevent them rusting. Wipe wooden handles with linseed oil when new and occasionally there-after, to prevent them drying out.

c Transport tools under vehicle seats or in a trailer or roof box to prevent accidents. Wrap edged tools in sacking or provide individual guards, both for safety and to prevent them damaging each other.

d Clean and lubricate hand winches with gear oil at the end of the task.

SHARPENING EDGED TOOLS

This covers the sharpening of tools in the field. For further details on filing and grinding, saw maintenance and replacing hafts and handles, see 'Woodlands' (BTCV, 1980).

a Sharpen tools at least twice a day when in use, or more often as necessary. Grass hooks and scythes need very frequent sharpening, and a quick touch-up every ten minutes is not excessive.

b Carry the correct whetstone for the tool. Fine cylindrical stones are needed for grass hooks and scythes. These and canoe-shaped stones can be used on billhooks and slashers. Canoe-shaped or flat rectangular stones are best for axes. Oval axe-stones can be dangerous as they are difficult to hold.

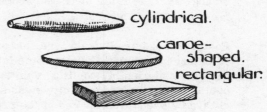

cylindrical.

canoe-shaped.

rectangular.

c Stones are fragile. Wrap them and carry separately from the tools.

d Always wear a glove on the hand holding the

sharpening stone. Place the tool on a firm surface such as stump, with the edge projecting. Spit on the stone to moisten it, as a stone used dry is quickly worn away. Hold the stone at an angle conforming to the existing taper of the blade. If using a combination stone, use the coarse side first to eliminate any flaws and give an edge, and then use the fine side to give a good polish and even taper. Sharpen with small circular motions. Take particular care to sharpen the hooked part of the billhook, as this does most of the cutting work.

e   Do not touch the blade to see if it is sharp. Sight along the edge, and you should see a uniform taper with no light reflected from the edge. Any reflected light indicates a dull spot which needs further sharpening.

Organising Group Work

a   Volunteers will work best if someone responsible for the site or path is present at the beginning of the day to explain the purpose of the task, and to work alongside them.

b   Before starting work, the leader should explain the use and care of tools, and set the day's objectives. Where a task seems dishearteningly big, it helps to see at the end of the day that a measureable amount has been achieved.

c   If work is being concentrated on one small part of a site, such as steps on a path leading to a waterfall, ensure the volunteers see the rest of the site early on during the task, so they can appreciate the context of the work they are doing. Walkers will inevitably get into conversation with volunteers, and the leader should ensure that volunteers are well informed about the site and the need for management, both for their own benefit and for explaining to others.

d   Footpath tasks are those where conservation volunteers are most likely to come into contact, and possibly conflict, with the public. Warning and publicity signs should always be used at either end of the path if it is open to the public. These signs are well worth a little time and trouble to make, and can be beneficially accompanied by a contributions hat or wellie.

e   The major problem with organising footpath

work follows from its linear nature. The task may involve a series of discrete operations such as stile building or cross-drain construction. This will require teams of two or three people working at each location, with each team having a set of tools and working independently. Supervision becomes difficult, and each team must know how to do the job.

Other work such as step building or boardwalk construction can usually proceed from the start only. This requires careful division of labour so that the ground preparation, carrying of materials and construction are kept running smoothly without delays and bottlenecks. In some situations it may be easier if, for example, everyone spends the first day carrying materials to prevent the line of communication becoming too stretched.

Vegetation clearance, and some types of surfacing such as stone pitching, may be worked in several places at once, each extending to join up. Ensure that this does in fact happen by the end of the task!

Many types of work can use a 'leapfrog' system, which allows small groups to work independently without becoming cut off.

f   Space is often very limited on paths, and the land on either side may be too steep or overgrown to be used. Always plan and instruct clearly where materials are to be dumped, especially if ground has to be excavated. It is very easy to get in a mess with materials and debris being trampled everywhere. If possible, pile materials onto tarpaulins or plastic sheets.

g   Few jobs will need a large group working together, and the leader must use his or her judgement in dividing the group, and swapping people around as necessary to keep everyone happy. Try to put new workers with an experienced person.

6 Path Clearance

This chapter is mainly concerned with paths through woodlands, copses and scrub, and paths enclosed by hedges through agricultural and residential land. Conditions vary greatly from one part of the country to another, and sheltered south and western areas will have much lusher growth than exposed or upland areas.

Reasons for doing clearance work may include:

a   To make the path easy to follow and pleasant to use. Increased use will keep the path open and reduce the need for future work.

b   To keep a right of way in use. By clearing a path and walking it, one is decreasing the likelihood of there being a successful order (England and Wales) to close it.

c   To improve the quality of the path and its edges. Clearing overgrowth allows wind and sun to dry muddy paths, and if done with care can be of benefit both visually and ecologically. Woodland and scrub edges with a gradation of vegetation are valuable habitats.

d   To channel use onto a particular route, and divert use away from ecologically sensitive areas.

Planning the Work

The statutory responsibility for maintaining rights of way is explained in Chapter 2. Permission must be gained from the landowner or local authority.

a   Consult the definitive map (see p11), and make sure all parties agree about the exact line of the path. If possible walk it with the landowner or occupier, but bear in mind they may be misinformed, or may even deliberately mislead. Any discrepancy should be referred back to the local authority. Removal of vegetation not directly in the line of the right of way could constitute an act of trespass for which damages could be claimed by the landowner.

b   There is no general right to go into a field adjoining a path to do clearance or to burn debris. If debris cannot be disposed of along the line of the path, come to an arrangement with the landowner about its disposal.

c   Unless clearance of all paths in a particular area is being undertaken, it is best if volunteer energies are directed at paths that will be used and so kept open. This is more likely on paths that make a circular route, a link, or lead to a viewpoint or feature of interest. Use will be greatly increased if the path is included in a printed guide. Liaise with the local authority over the priority of paths to be cleared, to fit in with any clearance or waymarking plans they may have.

d   Try to establish what will be done once the initial clearance is finished. It is very discouraging for volunteers to return to a cleared path only to find it overgrown again. The landowner or local authority may be able to maintain a path once it is cleared, or a local group or individual may agree to look after a path.

TIME OF YEAR

a   Ideally, lush non-woody growth should be cut back twice each year, the first cut being in May or June, and the second in August or September. In practice, most paths will be lucky to receive an annual cut, which is probably best done in June.

b   Woody growth, scrub, hedgerows and brambles should not be cleared during the bird nesting season from the beginning of April until the end of August. October and November are often the best months for major clearance jobs as working conditions should still be pleasant with the ground not yet at its wettest.

HOW MUCH CLEARANCE?

There is no statutory minimum width for a public right of way, and the width of any particular path will depend on local custom, and the terms of the dedication, if any (see p11). If the path has been created or diverted by a public path order, the width should normally have been defined in the order. In some cases statements accompanying the definitive map of rights of way give the agreed width of path. Always consult the landowner or highway authority about the height and width required.

An appropriate working minimum is 2m wide by 2m high for a footpath, and 3m by 3m for a bridleway. However, advice from experienced clearance workers is always to cut back as much as possible, usually to the boundaries of the path. A path cut back to 3m width will stay open for twice as long as a 2m wide path. Some shrubs such as blackthorn and bramble will regrow, unless dug out or treated, even in the presence of quite heavy trampling. As detailed below, a wide path should develop a ground cover of turf, which is easier to manage than a shaded and often muddy path.

Before starting clearance consider the following:

a   What sort of use will the path receive? Will it be by groups or individuals, one-way or two-way use? A newly opened path in a country park may attract a high level of use and will need to be at least 3m wide.

b   By making a bridleway wider than 3m you may attract illegal vehicular use. Likewise, by clearing a footpath wider and higher than is necessary you may attract horseriders. Bridlegates and barriers can be erected, but often the resources are not available. A fortuitous fallen tree which stops horses but not walkers may be useful to deter unwanted use. A tunnel of scrub 2m high will also discourage riders.

c   If time is limited, try to gauge the work so that the clearance is completed in one task or session. It may be better to clear right through to a narrow width and widen it later if there is time. A 'no through road' will only cause confusion and trespass.

Vegetation and shade

The width of a path through woodland or scrub obviously affects the amount of shade that is cast, which in turn determines the type of vegetation that will grow.

A path completely enclosed by scrub or low trees will have virtually no ground flora at all. Unless there is good natural drainage the path is likely to be often muddy as sun and wind cannot dry it out. Once cleared of branches which block it, the path is likely to stay clear for some time as non-woody growth will not flourish in the dense shade.

Ground under tall or open canopy will have a woodland flora of plants such as dog's mercury, bluebell, red campion and ground ivy, which are not resistant to trampling. (Beech woods are an exception in having very little ground flora due to the deep leaf litter, and the early emergence of the leaves in spring which shade out other plants.) The woodland flora can rapidly grow to obscure or block a path in the absence of trampling.

A path not enclosed by a canopy and receiving plenty of light will develop a flora mainly of grasses, especially in the presence of trampling or grazing. The width at which such a flora develops depends on the height of the trees and the alignment of the path. A path running on a north-south axis will receive more light than one on an east-west axis.

If it is feasible to clear to a width of about 4m, an attractive and relatively hard-wearing path can be made which is easy to maintain by machine. The initial clearance is likely to produce a flush of growth of weedy species such as, in the south, thistle, rose-bay willow herb and nettle, but these will decline with regular cutting, grazing or trampling to produce a grassy turf.

LANDSCAPING THE PATH

Often clearance will simply be a matter of cutting back growth to the original edges of a path, enclosed by banks or walls. On other paths there may be opportunity to vary the amount of clearance to increase visual and ecological interest.

a   Grade the edges by cutting at different heights or frequencies to create a diversity of habitats.

annual   2-3 years   5-10 years

PATH.

b   Clear glades at intervals along the path, especially on the northern side of the path where they will get maximum sunshine. These glades will attract certain types of butterflies. They will also provide pleasant places for walkers to rest.

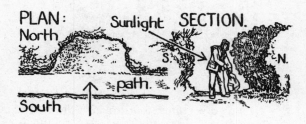

c   Cut sections in rotation each year to maintain a succession of habitat development.

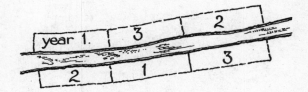

d   Vary the width to give visual interest, for example retaining a narrow section to emphasise the drama of a viewpoint ahead.

Narrow or winding sections can be created when clearing along old railway lines, to add variety to the path.

e   Views can be opened up by clearing scrub or trees, but consider first any unwelcome effects these may have. Views of nearby features such as a pond or stream may attract trespass by walkers on the path. Clearance of vegetation on a slope to open a view may expose the slope to erosion by rainfall, and encourage people onto the slope which will further hasten damage.

Viewpoints through tall trees are best opened up by cutting back the lower branches so that the canopy remains to protect the slope and to shade out growth beneath. View-

points made by lopping trees or clearing scrub will only be temporary features, unless they are maintained frequently.

Rights of way, although private property, attract public interest. There may be conflict, for example, between naturalists who would rather see overgrowth left undisturbed, and walkers and riders who want clear and easily passable paths. Adjacent landowners not responsible for the path may complain if vegetation is thought to be spreading weed seeds onto their land, or harbouring pests. Landscaping the path and being sensitive to these interests should help reduce conflict.

MARKING THE ROUTE

New paths through woodland or scrub need to be surveyed carefully to pick the best route (see Chapter 3 and 4). In very dense scrub it may be necessary to have two people, one with a tall pole for sighting the line of the path. A telescopic surveying staff is useful for this as otherwise negotiation of thickets will be difficult.

The route should be marked with stakes painted a bright colour, or by strips of fertiliser bag tied to branches. Shrubs, saplings or trees which you wish to retain can be tagged at the same time. Be consistent in the way you place the markers so that they mark either the centre or the edge of the path. It is usually better to put the markers to the edge so they can be left in place during clearance operations, and collected together at the end of the task for re-use.

Hand Clearance

ORGANISING A GROUP

This will depend on the type of vegetation and the number of volunteers. Space to work is usually very restricted, so if there are more than five volunteers, split them into groups with each taking a section, either 'leap frogging', or by dividing the total length by the number of groups. In very dense vegetation it may be difficult to work other than from one end of the path, but it should be possible for volunteers to follow the route taken by the person who marked the path. It is not unknown for one hardy volunteer to have to wriggle into the undergrowth and clear a space until there is room for another to follow!

With all clearance work, the main problem is in getting rid of the cut material. Sometimes it is possible to get rid of the debris off the edge of the path by dumping it in clearings, pushing it under overhanging scrub and filling holes in the ground. All non-woody material will rapidly bulk down and decompose. In this type of vegetation each person can cut and dispose of material. In clearance of thickets and woody growth there will have to be a division of labour between people cutting and people carrying and disposing of debris.

A possible division of labour is as follows. One person 'breaks the trail' using a slasher or billhook on brambles and light scrub. Material is then pulled back from behind him by one or two people using pitchforks or rakes, and carried to the point where it is to be dumped or burnt. They are followed by a person with a bow saw to fell any small trees or lop overhanging branches, and a fifth person to remove or treat stumps, and to do a final tidy-up with a pitchfork or rake.

Avoid having a bonfire if there is any other satisfactory way of getting rid of the material. Bonfires are enjoyable, especially in cold weather, but can take up to half the volunteer effort because of the time involved in carrying and dragging debris to the fire and in tending it (see p56).

On a path requiring only light clearance it is usually possible for each person or two sharing to have a set of tools such as a slasher, loppers and bow saw, and work on their own section of path.

NON-WOODY GROWTH

<u>Edging up and trimming</u>

Paths that are designed to have hard-wearing all-weather surfaces usually need their edges trimmed back every couple of years, in order to keep the surface free of vegetation.

1   Cut back overhanging growth with a grass hook or swipe. Use a crooked stick to hold the vegetation for cutting with a hook.

2   Using a mattock, cut through the turf along the line of the edge of the path, without cutting into the surfacing itself.

3   Shovel up the edgings, taking care not to scrape up too much of the surfacing, and either scatter off the path, or collect in a bucket or wheelbarrow for disposal or re-use. Edgings can be used as small turves to repair damaged sections of grass.

As a guide to rates of work, 1km of path was cleared in this way in June in about 30 man/days.

Grasses and herbaceous plants can also be cut with a scythe or swipe (see p 42). Scythes require some practice to handle effectively but are useful on wide paths or for cutting glades. Swipes are very easy to use, but are slower, and do not give such a clean cut through long grass. They are however much more versatile and can be used on nettles, young brambles and light branches. They are not robust enough for major clearance jobs, but are ideal for tidying and trimming.

<u>Brambles</u>

Brambles need to be attacked in orderly fashion from 'inside-out', as the outer growth is springy and rather resistant to cutting.

a   Using a slasher or grass hook, make two vertical cuts as shown, and one underneath to cut through any rooting stolons. The

mass of bramble can then be pulled away.
Bramble frequently grows along fence lines,
and if agreed by the landowner, remove the
bramble from both sides of the fence or it
will quickly regrow.

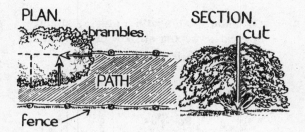

PLAN.   brambles.   SECTION.   cut
PATH
fence

b   Alternatively, for those not adept at using
     slashers, it is possible but slower to cut
     your way with loppers to the root of the
     plant, and then pull the growth away from
     the root end. You should then have a big
     bundle of growth which you can easily drag
     away. If you start cutting from the outside
     of the plant you will end up with lots of
     small pieces of prickly growth which are
     difficult to handle.

c   Bramble spreads by stolons, which are
     stems that bend down and root at the peri-
     meter of the plant. Sometimes a long stolon
     can be pulled away which is rooting at
     several points along its length. Always try
     to remove as many roots as possible, or
     else the plant will quickly sprout new growth.
     Herbicide treatment may be necessary (see
     p57).

Nettles

The stinging nettle (Urtica dioica), is a perennial
plant with a very tough and dense root system
which is difficult to remove from a path by digging.
The plant is easily cut or trampled but rapidly
grows again, and in the south, has a very long
growing season. Cutting must be frequent if no
other method of control is used. Scythes and
swipes are better than grass hooks for cutting
tall nettles as they allow one to keep a comfortable distance from the plant. Repeated cutting
and trampling will weaken the plant, but herbicide
treatment is usually advisable.

Bracken

Bracken (Pteridium aquilinum) is a fern, being
flowerless and reproducing by spores instead of
seeds. It spreads mainly by tough underground
stems called rhizomes.

Bracken is easily trampled or cut in spring and

early summer, but can present a formidable
barrier by August, especially on rich moist soils
in the south. If cutting is to be the only treatment,
timing is very important as it should be done when
the food store in the rhizomes is at its weakest.
The best method is to cut in mid June, and then
again at the end of July to remove any secondary
growth which will have further depleted the food
store. If only one cut is possible, do it at the
end of July. By August it is getting too late, as
sugars produced in the fronds are by then moving
down into the rhizome. Cutting tends to increase
the frond density, but fronds are short and weak
and the path will be easy to see and follow.

Bracken is not destroyed by trampling until the
level of trampling is such that other plants are
destroyed. Herbicide treatment is described on
page 58.

Other plants

Japanese knotweed (Polygonum cuspidatum) is an
introduced plant with extremely vigorous growth
and extensive underground stems. It has heart-
shaped leaves and a cane-like stem that grows up
to 2m tall and hardens by mid-summer. The stems
remain over winter. It is very difficult to control
and needs repeated cutting and treatment of the
young growth with glyphosate (see p57). It is
present throughout Britain and grows particularly
rampantly in the south and west, and should be
eradicated from paths without compunction.
Warn, and enlist the assistance of adjacent landowners as it spreads extremely rapidly!

Various species of thistle are common on paths
in agricultural areas. Cut plants down in mid
summer before they seed. Rose-bay willow herb
(Epilobium angustifolium) is a rapid coloniser of
disturbed ground and it may be wise to cut it in
flower to prevent it seeding. Foxglove (Digitalis
purpurea) and red campion (Silene dioica) may
flourish where a woodland canopy has been
recently removed, but decline in the following
years. They can be easily pulled up by hand if
necessary.

creamy white Flowers.

Rose-bay Willow Herb.

Japanese Knotweed.

All the plants mentioned above, except Japanese knotweed, have their own value in the British flora, as well as being food plants or egg-laying sites for invertebrates. Do not cut more than is necessary to keep the path clear, unless edges or glades are being managed as described above.

SCRUB

This term covers shrubs and small trees such as blackthorn (Prunus spinosa), hawthorn (Crataegus spp), elder (Sambucus nigra), holly (Ilex aquifolium), rhododendron (Rhododendron ponticum) and hazel (Corylus avellana).

The bow saw is the safest and most efficient tool for felling, with the billhook and pruners useful for trimming and snedding. Beware of blackthorn, as wounds caused by the thorns can go septic.

1   On bushy or overhanging growth such as blackthorn or holly you will probably need to cut away the lower branches with loppers to gain access to the main stem or trunk. Give yourself plenty of room, or you will restrict the length of saw stroke you can make.

2   If the stump is to be chemically treated, cut it as low to the ground as possible. If it is going to be dug or winched out, leave about 1m of stem for leverage.

3   Cut level or at a slight angle in the direction of fall, using the full length of the blade. A slight rocking motion, as shown below, gives greatest speed. Use both hands on a D-shaped saw until it is necessary to steady the tree. The triangular saw can be use one-handed. Use equal force on push and pull strokes to prevent the blade twisting.

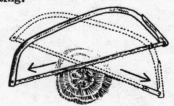

4   Steady the trunk as it starts to settle or move, and saw quickly through the last bit to prevent the stem splitting.

Billhooks are useful on multi-stemmed or coppiced shrubs such as hazel.

1   Cut away with loppers or secateurs any

young or springy shoots which the billhook may catch on.

2   Use the billhook one handed, and keep the other hand as high up the stem as possible, for safety.

3   Cut small stems with a single, slightly upward-sweeping stroke. Larger stems can be cut by notching.

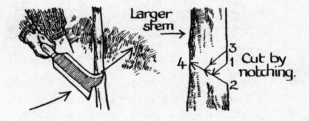

Stump removal

Removing the stumps of scrub has the following advantages:

a   For most species there will be no re-growth. Poplar can re-grow from roots left in the soil.

b   Any hazard to walkers is removed.

However, there are disadvantages:

a   Removal disrupts the surface and foundation of the path, and may hasten erosion on slopes.

b   Removal can be difficult and time-consuming, especially in rocky or compacted ground.

Stumps can be treated chemically to prevent or discourage regrowth, but does not solve the problem of the hazard to walkers. The following guidelines are suggested:

a   Remove all stumps from the line of a path which is going to be surfaced for intensive use. In this case a certain amount of ground disturbance will happen anyway.

b   Remove all stumps of species that have a low susceptibility to chemical treatment. This mainly applies to rhododendron, hawthorn and sycamore.

c   Remove all stumps from paths or edges that are to be maintained with grass-cutting machines or hand tools, to prevent damage to them.

d   If time is short or the path is in a remote location which only well-shod walkers should reach, treat all stumps chemically.

To remove stumps by hand:

1   Dig with a spade or mattock around the base of the stumps to expose as many roots as possible.

2   Chop through the roots with a grubbing mattock. Do not use an axe or you will blunt it on any earth or stones around the roots. Lever under cut roots with the other end of the mattock to loosen them.

3   Try and loosen the roots by levering on the stump. Chop under the stump with a sharp mattock or spade, and lever out using a crowbar or Tirfor winch.

4   Fill the hole, using material dug from off the path if necessary, and tread well down to leave an even surface. Finish to match the existing surfacing. If possible, repair a grass path with turves.

Ecological value

The value of any particular plant depends on the balance of species in the habitat. Try to conserve any that are uncommon but typical of the habitat. If a lot of clearance has to be done, conserve a representative sample of the range of species in the habitat. Try and visit the path that has to be cleared in the summer before clearance takes place, when it is easier to identify and evaluate the species. Note down and mark any that you wish to save.

Some species can be treated in such a way as to retain some ecological value, while making them more amenable as path-side plants.

a   Trim holly bushes of their lower branches to make a standard 'tree'. This will look very unbalanced at first, but should grow into a reasonably shaped tree.

b   Ash, birch and other species can be encouraged to grow as single trees. From each stool, cut all stems but one, retaining the strongest and straightest. Growth will then concentrate into the single stem.

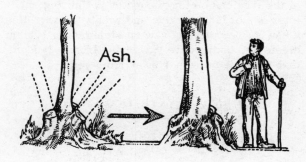

c   Although infected, elm appears to be not susceptible to the attack of the ambrosia beetle (responsible for Dutch Elm Disease) until it is about 3m high. As elm suckers rapidly it is possible to try and keep the plant healthy by continually cutting to restrict growth to this height. This maintains some scrub habitat for cover, and in particular conserves the habitat of the white-letter hairstreak butterfly, which is dependent on elm.

TREES

Unless a wide path is to be cleared through dense woodlands, the felling of mature trees is not usually required for footpath clearance. Felling of trees is described in 'Woodlands' (BTCV 1980).

If side branches must be removed, use a pruning or bow saw, with an extension to reach high branches. Always make the cut on the outer side of the branch bark ridge. A cut on the inner side will expose the trunk to fungal infection. Make the cut slanting slightly outwards.

If the branch is thicker than 30mm, it is likely to tear if only one cut is made. Make three cuts as shown.

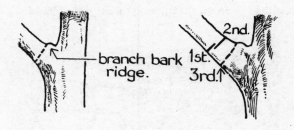

Pollards

Pollarded trees are valuable habitats, as well as being traditional features of hedge banks and boundaries that footpaths often follow. The life of the tree can be lengthened by pollarding, at intervals of between five and twenty years. Use a bow saw or pruning saw on stems up to 100mm diameter, taking care if working from a ladder that the ends fall clear away from you.

Work clockwise around tree, if right-handed.

PLAN

ladder

It is also possible to pollard existing trees, which may provide a useful compromise if the landowner wishes to clear trees to allow more light onto the path or adjacent field. This is best done when the trunk is 100mm-150mm diameter, at a height of 2m-3m. Cut above the height which any browsing stock can reach, taking care not to split the trunk or infection may result. Follow standard tree felling practices for this and for maintaining old pollards which have stems over 100mm diameter.

Fallen trees

Large trees that have fallen across the path should be cut at either side, using a chain saw or cross-cut saw. Roll the freed section off the path, or cut it up for use.

Alternatively, the trunk can be left as a barrier against horse or motorbikes, if required. To cut a step, make two vertical cuts with the saw, slightly below the finished height of the step. The step should be not higher than 400mm, to be easily negotiable by walkers. Cut the step away using either the chain saw, or by knocking wedges in on each side. Knock them in by equal stages on either side so that a horizontal cut is made.

SECTION
wedge

400mm

Smaller trees only need one cut, and then the top can be dragged off the path. Cutting up and disposal of fallen trees depends on the habitat and the situation. You may be able to use the timber elsewhere along the path for steps or barriers, but if not, leave it in place as the rotting timber provides a good habitat.

Saplings

By way of balancing the destructive element of path clearance work, a positive contribution to the landscape can be made by tagging saplings. This involves attaching special tags, obtainable from the Tree Council, to saplings which are to be left to grow into mature trees. The campaign was started to help the replacement of roadside and hedgerow trees, and the tags are a bright colour to be easily visible to machine operators. The same system can be used in path-side copses and hedges to be followed during future clearance work. Tagging must be done in agreement with the landowner if the system is to be respected. Choose straight, healthy saplings, leaving at least 10m between each. Further details on tagging are available from the Tree Council.

Machine Clearance

Excavator

Clearance of new paths through scrub or woodland can be done by a mini-excavator, provided there is suitable access for the machine. A $3\frac{1}{2}$ ton excavator, with a bucket and a cutter bar for clearing vegetation, is suitable. Such machines are very manoeuvrable, and can cut winding paths to a minimum $2\frac{1}{2}$ metres width. The excavator should be followed by a dumper truck into which the bucket is emptied, and the material taken as necessary to build up sections of path elsewhere.

As an example, 800 metres of path were cleared in four days through thick scrub and undergrowth at Maple Lodge Lake, Denham. All clearance work was done before the volunteers arrived to do the surfacing, so there was no risk of accident to them. The machine must only be driven by a trained operator. Firms are listed under 'Contractors' Plant and Machinery Hire' in the Yellow Pages Telephone Directory.

Tractor-mounted machines

Although these are outside the experience of most voluntary path workers, tractor-mounted machines can be invaluable to those people responsible for the maintenance of bridleways and byways, especially in the south and west where vegetation growth is rapid.

Flails, either as hedgers mounted on booms (p181 6.1), or as front or rear-mounted mowers, are the most useful type of machine and can deal with rough grass and woody growth. If the flails are sharp and the correct type is used for the job, the result is not necessarily unattractive, and an annual 'trim' need not be destructive to wildlife. In the absence of more labour intensive methods, it may be the only way of maintaining a path.

Machines may be available either through the local authority, or from an agricultural contractor. Flail mowers for 'compact' tractors are now manufactured, with a cutting width of 1.3m, for use in confined spaces (p181 6.2, 6.3). For more extensive scrub clearance operations, a heavy duty flail mower and mulching machine can be used (p181 6.3).

Pedestrian flail mowers

These operate on the same principle as the tractor mounted machines, with short blades fixed to a rotor so that they pivot out of the way if an obstruction is met. Grass and woody growth up to about 2 years old is pulverised to a mulch, and raking up and disposal is not necessary.

Flail mowers should only be used by trained operators, and in very dry, dusty vegetation a mask should be worn over the mouth and nose. Slopes of up to about 15 degrees can be cut. Machines are manufactured with cutting widths of 563mm, 635mm and 747mm (p181 6.4, 6.5).

Brush cutter

This is a portable machine with a rotating cutting blade on a long arm. The weight of the machine is taken by a harness or shoulder strap. It is very suitable for path clearance as it can be transported in an estate car, carried over stiles, and used in confined spaces and on any type of ground. The machine is supplied with a selection of blades for cutting vegetation up to scrub and small trees. Most also have a nylon thread 'strimmer' head for grass cutting.

The choice of engine capacity will depend on the type of vegetation likely to be encountered, the amount of use it may receive, and the stamina of the operators. For most groups a 37cc (or nearest) model is probably suitable, but for lighter use a 23cc model may be preferred. To date, the Husqvarna brush cutters have proved

popular as they are easy to start, reliable and not too noisy. However, there are likely to be new makes and models coming onto the market, so it is as well to shop around (p181 6.6 - 6.9).

Note the following points on safety:

a   Goggles must always be worn by the operator. A helmet with face mask and ear defenders is recommended.

b   Boots with steel toe-caps must always be worn by the operator.

c   Any other workers should keep at least 10m distance from the operator as there is a danger of flying stones and splinters.

Brush cutters are best used by teams of two or three people, with one cutting and the other one or two raking and disposing of debris. They are not necessarily the answer for voluntary groups, as in common with most other machines, the remainder of the group tend to be left with only the clearing up work, which can be monotonous. Any group considering purchase should also first check that they have a member willing and able to maintain the machine.

Knapsack mower

This is a similar type of machine to the brush cutter, but the motor is carried knapsack style, thus making the machine less tiring to use (p181 6.10).

Disposal

Try and do as little carrying and burning as possible, while leaving the path completely clear of debris, and its edges attractive.

Non-woody growth rapidly rots down and should cause little problem. Brambles and scrub are more awkward as they are bulky and difficult to 'lose'. Consider using branches and scrub for blocking off short cuts and gaps in hedges, or for covering bare areas to stop trampling and allow vegetation recovery. A few piles of brushwood in odd corners are useful habitats for over-wintering and nesting wildlife. Large amounts of scrub and timber will have to be burnt or removed for disposal.

Snedding

Most scrub will have to be snedded both for

stacking in a trailer or for efficient burning.
Before snedding, drag the tree or shrub to the
point where it will be burnt or stacked, as this is
easier than carrying bundles of cut material.
Using a snedding axe, billhook or loppers, remove
all side branches, starting at the base of the stem.
If using an axe or billhook, stand on the opposite
side to the branch you are cutting, to protect
your legs.

BURNING

Burning is time-consuming so avoid it if you can.
If not, try and get permission from the landowner
to burn on adjoining land as space on the path is
likely to be restricted. It may be necessary to cut
a temporary gap through which to drag the material. This can be blocked afterwards with cut
scrub.

In choosing the site, consider the following:

a   Avoid overhanging trees and interesting
     ground flora, and keep the fire well away
     from the trunks of trees as heat can damage
     far beyond the visible flames. Smooth
     barked trees such as beech, sycamore and
     ash are especially susceptible to scorching.

b   The fire will enrich the soil with mineral
     ash which encourages the growth of thistles
     and nettles. This effect can be reduced,
     but not completely prevented, by shovelling
     up the wood ash when the fire is dead, and
     removing it.

c   Unless the path is going to be surfaced,
     avoid having the fire on the line of the path,
     as both the heat and the trampling will
     damage the soil structure and make the site
     liable to be muddy in the future. If possible,
     choose a slight knoll beside the path.

d   Try and site the fire in as central a location
     as possible to minimise carrying. If the
     scrub is so thick that you can only approach
     from one end, you will probably need
     several fires. To save time, you can carry
     hot embers in a metal bucket or shovel from
     the dying fire onto the next site to get the

fire going quickly. It is also possible to
burn on a metal sheet, such as corrugated
iron. This can then be dragged, in suitably
flat terrain, from one site to another. This
speeds fire lighting and avoids damage to
the ground.

e   Alert the police and fire service in advance
     on sites of high fire risk, such as heath-
     lands and conifer plantations.

Starting and tending a fire

Getting a good fire going is not as simple as it
seems, and is best done by an experienced person.
Even when the fire is going well, it should be
constantly tended by one person to keep it burning
efficiently.

1   Even the most optimistic fire lighter will
     need matches and dry paper, but usually
     dry kindling, sump oil and a tyre or inner
     tube should be taken as well. Paraffin and
     diesel are useless, and petrol very danger-
     ous, as fire-lighters. Try and get a small
     tyre such as a motor-cycle tyre if you have
     to carry it some way. The disadvantage of
     tyres is that they give off noxious fumes
     while burning, and most contain reinforcing
     wire which must be cleared up afterwards.

2   Stuff the tyre with paper and pour over some
     sump oil. Build a stack of dry wood about
     a metre high over the tyre to take advantage
     of the heat produced. Light the paper.

3   Build up the fire using thin dry wood which
     lies close together. Place larger branches
     on only when the fire has a good hold.
     Always put them on with their butt ends to
     the wind, which allows a free flow of air
     into the heart of the fire.

4   Blackthorn, hawthorn and old bramble stems
     are particularly awkward to burn because
     they do not pack down well. Allow time to
     sned them, and if possible put them on the
     fire with other material that burns well.

5   If you are returning to do more work the
     next day, pile up the embers to keep them
     hot so the fire can be quickly re-started.
     At the end of the task, clear up and restore
     the fire site by removing any metal such as
     reinforcing wire, and scattering any unburnt
     stumps or logs. Shovel up and remove as
     much wood ash as you can. Rake the site.

Herbicides

For ecological and visual reasons many groups
and individuals will prefer not to use herbicides.
However, their use will allow more efficient
clearance, and lengthen the period before repeat
clearance is necessary. Herbicides should only
be used by experienced volunteers.

SAFETY PRECAUTIONS

a   Keep herbicide concentrates in their
     containers. Read and follow the manufact-
     urer's advice.

b   Provide soap, clean water and towel for
     washing hands and other exposed skin in
     case of spillage, and for after work and
     before meals. As a second best use a
     waterless skin cleanser and paper towels.

c   Mix herbicide solutions in a workshop where
     there are washing facilities. Wear disposable
     plastic gloves and a face shield when mixing
     concentrates and wash immediately if you
     splash yourself. An eye bath must be
     available.

d   Wear suitable protective clothing at all times.
     For stump treatment, knee length wellington
     boots and chemical resistant gloves are
     essential, and waterproof leggings are
     advisable. For spraying, wellington boots,
     chemical resistant gloves, waterproof over-
     trousers and jacket are essential, and a face
     shield and ori-nasal mask are recommended.
     Take care not to come into contact with
     contaminated parts of protective clothing
     when taking it off.

e   Take the herbicide solution to the work site
     already mixed, either in the sprayer, or in
     plastic screw-top containers, clearly
     marked, for stump treatment.

f   Never leave herbicides on the path overnight.

g   Avoid contamination of waterways.

h   Use sprayers only in calm weather to reduce
     the risk of chemical drift.

i   Store herbicides and protective clothing
     away from foodstuffs and out of the reach
     of children and animals.

j   If a person who has been using chemicals
     falls ill or is supected to have ingested some
     chemical, give first aid and call a doctor

immediately, or take the person to hospital.
Ensure the doctor knows which chemical
is suspected.

Disposal of containers

Some local authorities will collect empty herbicide containers, which should have been rinsed
out. Tip the rinsing water onto bare earth to be
absorbed, not into a drain or ditch. Otherwise,
bury all containers at least 500mm deep, well
away from ponds and watercourses. More
detailed advice is contained in the MAFF leaflet
'Code of practice for the Disposal of Unwanted
Pesticides and Containers on Farms and Holdings'.

Instructions for use

Herbicides must always be used in strict accord-
ance with the manufacturer's instructions. Each
of the manufacturers listed in Appendix B issue
detailed information on application rates,
susceptibility of different species, and optimum
timing of treatments. This information is available direct from the manufacturers, along with
the addresses of local suppliers.

Manufacturer's information should always be
studied carefully, not only for safety reasons,
but also for the most effective and economical
use of the product.

NON-WOODY GROWTH

Nettles, brambles and other non-woody growth
can be sprayed with a translocated herbicide
such as glyphosate (sold as Round-Up or Tumbleweed) or triclopyr (Garlon 2). Round-Up is a
more active formulation than Tumbleweed, and
although more economical, requires greater care
in its application (p181 6.11).

The most effective use of these herbicides in
path clearance is on the secondary growth of
vegetation which has been cut about four weeks
previously. Food reserves of the plant are
weakened by the initial cutting, and it is easier
to reach and treat effectively the new growth,
than a tall, dense mass of vegetation. This also
avoids the problem of leaving large stands of
brown and dying vegetation, which looks very
unsightly. Round-Up can also be used for complete
control of growth along fences where neighbouring
landowners complain of weed spread.

If there is no time to cut the vegetation and wait
for re-growth before spraying, then simply spray
the unwanted growth in April or May.

Round-Up or Garlon 2 should be applied by knapsack sprayer. These are described in 'Woodlands' (BTCV 1980). Ultra-low volume sprayers are not recommended for non-selective herbicides, such as Round-Up, as the minute droplets are quickly dispersed in the air, and may reach outside the intended target area. Spray only on a dry day.

Round-Up can also be applied by a special glove with a pad into which the herbicide is drip-fed from a waist pouch. Individual plants can then be treated without risk of contamination to surrounding vegetation.

Bracken

Bracken should be sprayed with asulam (sold as Asulox - see p181 6.13) at the stage when the fronds are fully expanded but not lignified. There should follow a 95% reduction in the number of fronds the next year, but unless trampling is heavy, the remainder will re-invade over the next four or five years. Spot treatment of re-invading fronds in the second year after the initial treatment, combined with an increased level of trampling, should give sufficient control.

Use a knapsack sprayer, and spray only on a dry day. Do not cut the treated bracken for at least four weeks, to allow translocation of the herbicide to the rhizomes.

STUMPS

Stumps of trees and shrubs can be treated to prevent or reduce re-growth with either ammonium sulphanate (sold as Amcide - see p181 6.14), or triclopyr (sold as Garlon 2 - see p181 6.12). The procedure for treatment is as follows:

1   Mix the herbicide solution according to the manufacturer's instructions. Different rates of dilution are given for Garlon 2 according to the species being treated, so ensure that the correct dilution is used.

2   Cut the stump as low as possible to minimise the area which needs treating.

3   Empty sufficient herbicide for one application into an open container, and paint it on the stump and basal bark, to the point of run-off. This should be done as soon as possible after the stump is cut, as healing processes which seal the stump begin immediately it is cut. Alternatively, the solution can be applied using a 'foliar

feed' hand sprayer, available from garden centres. Label the sprayer with a permanent marker, and do not use for any other purpose.

Amcide can also be applied as crystals, which are placed along a V-shaped notch in the stump.

WOODY GROWTH

Fosamine ammonium (sold as Krenite -see p181 6.15) is a woody growth regulant. If applied in autumn, it is absorbed by the foliage, buds and stems, but there is virtually no visible effect at the time of treatment. Autumn leaf fall occurs normally, but in the spring treated buds fail to open. If the whole shrub or tree is sprayed it will die within a year or two, depending on the species. It is not translocated, and therefore can be used on selected branches or on one side of the tree only, to prevent new growth. Its main advantage along paths is that there is no obvious sign of treatment, and die-back appears 'natural'. It has little effect on underlying vegetation, and is not a hazard to wildlife.

7 Drainage

Poor drainage is the cause of most problems in path management. A typical situation on a path enclosed by hedges or walls is that surface and sub-surface water collects and causes water-logging and puddles of standing water. Trampling compacts the wet ground, further impeding drainage. Waterlogging prevents the growth of grasses and other vegetation which would help protect the surface. The path becomes a morass of mud in wet weather, and even in dry conditions has a pitted, uncomfortable surface on which to walk.

On unenclosed paths walkers simply avoid the muddy patch or puddle, so spreading the trampling and destruction of vegetation and soil structure, until a wide area may be affected. On enclosed paths walkers may make new paths along hedge banks to avoid the mud, or trespass on adjoining land.

On slopes the damage is not just to vegetation and soil structure, but to the ground itself, which is washed away as water runs down the slope. A typical cycle of events is that vegetation on slopes is worn away by trampling, which exposes the thin soil to rapid erosion by water running down the path. This leaves a loose and slippery path, which is abandoned in favour of another line, that is then rapidly reduced to the same condition. This has a very damaging effect on the landscape.

Approaches to the Problem

Consider in order each of the following approaches, and decide which are most relevant to the problem you are trying to solve.

1   Is there an existing drain which is blocked? Many paths were properly drained in the past, and can be quickly and sometimes spectacularly improved by locating the old drain, and unblocking it.

2   Investigate the use which the path receives. It may be that horses or wheeled traffic are using the path illegally, breaking piped culverts and the edges of open drains, as well as destroying vegetation and soil structure.

3   It may be easier in the long run to avoid the problem by re-routing the path. The new route must be chosen with care to ensure that it will not develop a similar problem. Re-routing rights of way may involve a diversion order (see p11).

4   Design and construct a new drainage system, as described below. Note that a camber or cross-fall (see p75) to shed water off the path is the simplest form of drainage.

5   Lay surfacing material on the path. This may be effective either in providing a strong surface which cannot be cut up and eroded when wet, or by actually lifting the level of the path above the water table. See Chapter 8.

6   Construct a boardwalk or bridge, which may avoid having to drain an ecologically interesting wet area. See Chapter 9.

Locating old drains

Paths likely to have drains are major routes between villages and settlements, and those showing other signs of construction such as walls, revetments and surfacing.

a   Visit the path in wet weather and find the point where water is running onto the path. Dig around at this point, and you may uncover a culvert opening.

b   Examine ditches alongside paths, especially after rainfall. A disturbance in the water, perhaps with muddy or rusty ochre colouring, can indicate a submerged drain. Staining on the side of the ditch may be a sign of a blocked outfall.

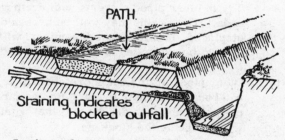

PATH

Staining indicates blocked outfall.

c   Look out for signs along the path such as holes in walls which no longer seem to have any function. Large stone slabs may indicate the top of an old culvert. In some areas, the outfalls of drains were traditionally marked by planting a single holly tree in the hedgerow.

Smoot Hole.

d   Muddy patches on the path may be the result of sediment dropped by water which previously flowed through a culvert. If scraping away some mud reveals a surfaced layer, there is a good chance of there being a blocked culvert underneath.

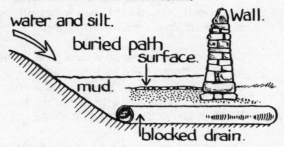

e   Talk with local farmers and residents who may remember where drains used to function.

f   Look up old maps and records, as these may show drainage systems. Old 6" to the mile maps should be obtainable through the local authority or archives office. Most cultivated land has been drained in the past, and large estates usually had records of any drainage work done.

g   Aerial photographs often reveal evidence of old drainage systems which show up as dark lines. Most areas of the country have been covered, and the local authority should have copies. If not, they are obtainable from commercial aerial photography firms.

Restoring old drains

a   Try to clear piped culverts with drainage rods. Tile drains may well be broken or displaced, in which case the culvert will have to be excavated and a new pipe laid.

b   Other piped drains are more difficult to rod because of their length. Rod as far as possible, and then dig down at that point to try and locate the blockage.

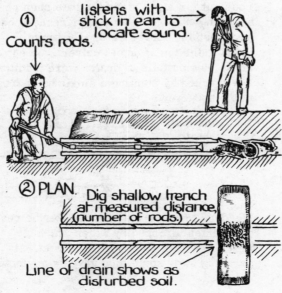

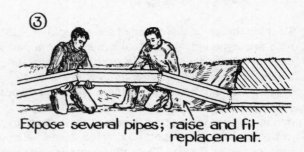

Expose several pipes; raise and fit replacement.

c   Blocked stone culverts will have to be dismantled and rebuilt. Rocks can easily wedge inside and block the culvert again if the sides are not smooth. It may be easier to install a plastic pipe inside. Build the stonework so that it hides the ends of the pipe (see p72).

d   Dig ditches as necessary at the infall and outfall of the culvert.

Designing Drainage Systems

DRAINAGE GLOSSARY

In most cases 'drains' can be either pipes or open ditches. A drain can act in any of the ways shown:-

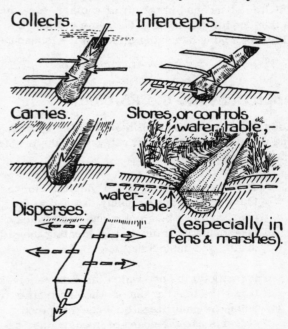

The diagrams below show two simple examples of drainage systems for paths.

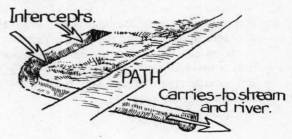

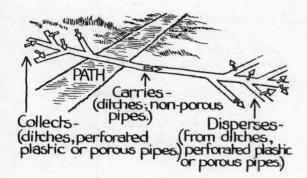

Collects – (ditches, perforated plastic or porous pipes.)

Carries – (ditches; non-porous pipes.)

Disperses – (from ditches, perforated plastic or porous pipes.)

The following terms are used in this book to describe various types of drains:

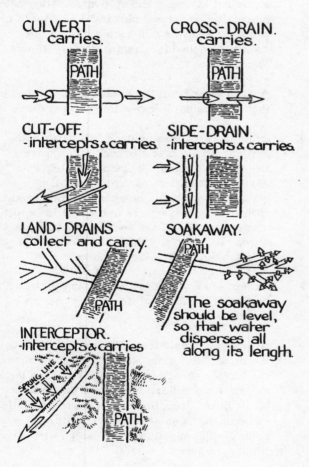

CULVERT carries.

CROSS-DRAIN. carries.

CUT-OFF. – intercepts & carries.

SIDE-DRAIN. – intercepts & carries.

LAND-DRAINS collect and carry.

SOAKAWAY.

The soakaway should be level, so that water disperses all along its length.

INTERCEPTOR. – intercepts & carries

SURVEYING THE SITE

a   Try and find out how the water reaches the path, and decide whether it is surface or sub-surface flow.

Surface flow in streams is usually fairly simple to deal with as these can be taken across or under the path. Surface flow which occurs as direct run-off of rainwater falling on slopes above the path is more problematical. Side or interceptor drains will be needed to divert the flow away from the path. Rainwater falling on the path

itself should be shed by a camber or cross-fall (see p75).

Direct run off.

Sub-surface flow can cause persistent water-logging throughout the year, and is often spread over a wide area making the problem difficult to deal with. Look out for places where the water comes to the surface, where flow can be intercepted and taken away from the path. This usually occurs at the foot of steep slopes or where solid rock is exposed.

Sub-surface flow.

b   Inspect the path in a variety of weather conditions if possible. Mark the places where storm-fed streams run onto the path. Use wooden pegs well hammered in at the edge of the path so they do not present a hazard. Number them for reference, and paint them a bright colour if they are likely to get lost in vegetation. Photographs will be useful for reference (see p16), as conditions may be very different when work is to be done.

c   Consider where the water will go once it is drained off the path. It is not enough to simply pass the problem on to the adjacent landowner. Consult the landowner for advice as it may be possible to put the water into an existing drainage system. If not, a soakaway may be needed. Ensure you are not going to cause a similar problem lower down on the path by unintentionally diverting the water back onto it.

d   It is possible to estimate the required size of pipe by calculating the run-off of water from a catchment area. For path drainage, this is probably only relevant for the design of large diameter culverts, or for carrier drains leading into another drainage system. The procedure is given on page 70.

MATERIALS AND METHODS

In many cases, only a simple drain will be needed, such as a cut-off or a culvert, and often these can be built with local materials. In other cases, the lie of the land and the source of the water may require a more complex drainage system.

Consider the following questions:

a   Should the drains be open ditches or pipes buried in trenches? Ditches are quicker and cheaper to construct, but require maintenance and may spoil the appearance of the area. Consider also whether ditches will act as barriers to keep people on the path, or whether they may be a hazard to walkers or riders. In very confined spaces there may not be room for a ditch, for example, alongside a path, and a pipe will have to be laid. Pipelaying involves more expense, but if properly done, provides an unobtrusive and low maintenance system.

b   Can a machine be used? For major schemes which involve the clearance of new paths, it may be worth using a machine, especially if this can also be used for clearance work. Ditch-digging by hand can be hard and monotonous work if the ground conditions are difficult, and so may not be done to the required standard.

DITCHING PROCEDURE

1   In most types of ground ditches need sloping sides for stability. A useful rule to remember is that, even in stable soils, a ditch must always have a top width at least twice that of the bottom. In many soils, three times the bottom width will be necessary. Angles of repose of various types of soil are given on page 21.

2   The gradient of the ditch will normally be dictated by the lie of the land. Minimum recommended gradients of about 1 in 400 can only be measured with surveying equipment, and are not relevant to most footpath tasks. More important is to keep the gradient of the ditch bed as even as possible, so that flow is not hindered. Avoid ditch gradients steeper than 1 in 30 or erosion of the ditch is likely to occur. Erosion of steep ditches can be reduced by building check dams of stone, timber or brushwood, at intervals across the ditch. These slow the flow of water, but require periodic cleaning out as silt accumulates.

3   Keep the ditch line as straight as possible. A line of sand sprinkled on the ground is easier than leaving pegs and lines in place, as these tend to get knocked.

4   Where a subsidiary ditch joins a larger ditch, make sure it enters at a gradual angle, to prevent erosion occuring at the junction. The bottom of the subsidiary ditch should be slightly above that of the main ditch.

5   The tools required will depend on the type of ground encountered. Spades may be sufficient in loams and cultivated soils, but more often forks and picks will be needed to break up stony or compacted ground. A rutter is needed for digging ditches in peat (see p43).

6   Always work from the lowest point uphill, so that you are not working in a flowing ditch.

7   Often the spoil can be used for filling holes, building up the path surface, or for restoring eroded ground. Work out the logistics carefully at the outset, so that spoil is moved the minimum number of times, and without unnecessary trampling across soft or wet ground. If the spoil is not being used, spread it on the downhill side of the ditch, or on the opposite side of the path.

8   Except when using a rutter, stand in the bottom of the ditch while digging, not on the sides.

9   Always work as a team along a length of the ditch. For example, one person strips the turf, a second loosens and removes the top spit, and a third cleans out the bottom of the ditch. If volunteers work individually on sections these are likely to differ in depth, width and alignment, which will hinder the flow of water.

LONGITUDINAL SECTION:
Order of digging.

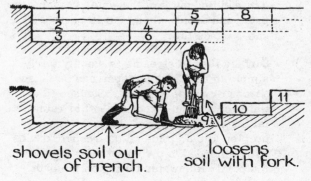

shovels soil out of trench.

loosens soil with fork.

PIPES AND PIPELAYING

The table below indicates the range of pipes available, and their suitability for path drainage work. An important distinction is between perforated (permeable) or unperforated pipes. Perforated pipes must be used wherever the drain is designed to collect or intercept water, or to disperse it back into the ground through a soakaway. Unperforated pipes must be used where the drain is simply carrying the water under the path or into another drainage system.

Apart from culverts over 225mm diameter, plastic (uPVC) pipe can be used for most path drainage work. Flexible uPVC pipe is inexpensive, light to handle and easy to install. It can be cut with a sharp knife or hacksaw, and joined with push-on connectors and junctions. Although care should be taken in laying, differential settlement is not usually a problem as the pipe is flexible enough to permit some soil movement without being damaged.

Bedding material

All pipes should be laid on a firm base to minimise the amount of settlement. Most clayey, silty and loamy soils give a firm enough base, but any dips or patches of soft ground should be filled with suitable bedding material, such as gravel or chippings of about 30mm diameter down to 5mm. Very rocky ground or peat will require a layer about 50mm deep of bedding material.

Backfill

In most situations, the efficiency of a permeable pipe in collecting water will be increased if it is covered with permeable backfill. This should be either washed gravel, stone chippings, slag or hard clinker of size 30mm down to 5mm. It must not be at all powdery, or it will block the holes in the pipe. If suitable material is available nearby then it is always worth using. If not, and material has to be bought and transported to the site, a decision will have to be made about

TABLE 7a   TYPES OF DRAINAGE PIPE AND THEIR USES

Type	Use	Sizes (diameter in mm)	Suppliers
Rigid uPVC unperforated	Seepage culverts	110, 160, 200.	Brett
Rigid uPVC perforated	Side drains, land drains, interceptors, soakaways.	54, 90, 110, 160, 200. 50, 70, 90.	Brett, Wavin.
Flexible unperforated	Seepage culverts	60, 80, 100, 125, 160, 200.	Wavin
Flexible perforated	Side drains, land drains, interceptors, soakaways.	60, 80, 100, 125, 160, 200, 310.	Wavin
Steel	Large culverts	300, 400, 500, 600, 700, 800, 900, 1000, 1100 etc to very large sizes.	Armco
Concrete	Large culverts	225, 300, 375, 450, 525, 600, etc up to 2400.	Rocla
Vitrified clay (impermeable)	Large culverts	400, 450, 500, 600, 700, 800.	Hepworth
Vitrified clay perforated	Side drains, interceptors, soakaways.	100, 150, 225, 300.	Hepworth
Clay channels	Cross drains, cut-offs.	100, 150, 225, 300	Hepworth

See page 181-182 (7.1-7.5) for addresses of manufacturers. Most firms supply technical details with advice on joining pipes, laying, etc.

whether or not it will be cost effective.

Backfill above permeable pipes should be firm but not compacted, or permeability will be reduced. The spoil from the trench should be mounded at the surface to allow for settlement. Permeable backfill can be used to the top of the trench, which then forms a French drain (see p68).

The requirement for backfill above impermeable pipes is quite different, as it should be stone-free and capable of being compacted so that it is absolutely firm. This is especially important for culverts. The suitability of the backfill can be tested as follows:

1   Fill a 250mm length of 150mm diameter pipe with a sample of the backfill. Level it off, remove the surplus, and empty it into bucket.

2   Using a quarter of the material at a time refill the pipe from the bucket, tamping down each addition to maximum compaction.

3   Measure the distance from the top of the pipe to the top of the compacted material. Divide this measurement by 250 to give the 'compaction factor'. If this figure exceeds 0.3, the material is unsuitable.

Pipelaying

1   Follow the procedure given for ditch digging, but dig the trench as narrow and steep sided as possible. The trench should be three times the diameter of the pipe. Make the bottom of the trench as neat and even as possible, removing rocks and tree roots and filling any holes with bedding material. Lay the pipe as soon as possible after the trench is dug.

2   If clay pipes are being used, lay these along the side of the trench.

3   Start laying the pipes from the outfall end, and work uphill. The Hepworth clay pipes are joined simply by pushing them together.

4   Inspect the alignment of the finished pipeline and adjust if necessary.

5   For impermeable pipes, backfill enough spoil to secure the pipe in position, up to about the midpoint of the pipe's diameter. Compact, then add more in 150mm layers, tamping down carefully. Mound to allow for settlement. Backfill permeable pipes with permeable backfill, not compacted.

ECOLOGICAL EFFECTS

Rights of way rarely cross large areas of wetland, and such paths as do are usually access paths on nature reserves, for which boardwalks should be built. However, many paths cross small patches of boggy ground, whose ecological value would be reduced if they were drained to improve the path. A decision over whether to drain will have to be made by considering how much similar habitat there is in the area. The priority though, is to ensure that the path or boardwalk will contain the trampling, as a wet habitat can be destroyed just as quickly by trampling as by drainage.

To avoid lowering the water table and so degrading the habitat, the path can be raised above it by using surfacing material over a porous membrane (see p78).

Crushed stone.

Water Table. Porous membrane.

Unfortunately, raising the path is often not possible as the material to do it is not available. In this case drains will have to be dug, or the area will only be damaged by trampling as people try to avoid the mud. To minimise effects on the habitat, do not dig land drains all across the area, but instead dig a side drain leading into a culvert or cross drain. Although this will lower the water table along the edge of the bog, a peat bog will hold water and the main area should be unaffected. The side drain also discourages walkers from stepping off the path. This system can be improved by putting an elm board along the drain, as shown. This maintains a high water table throughout the bog.

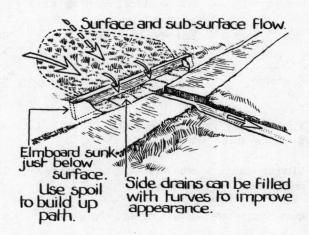

Surface and sub-surface flow.

Elmboard sunk just below surface.
Use spoil to build up path.

Side drains can be filled with turves to improve appearance.

The water table of wet areas not of peat can be kept high by constructing a simple sluice on the upper side of the cross drain or culvert. A side drain cut as shown in the preceding diagram would tend to collapse in waterlogged mineral (non-peaty) soils.

SECTION through Sluice.

Boggy areas that are already heavily trampled are likely to be of virtually no ecological value in that state. Dig as many ditches as necessary to get water off the area, which should slowly recover if trampling is successfully confined to the path.

Habitats can also be adversely affected by the raising of the water table. Do not place drains where they can cause waterlogging around trees.

Water-logging damages or kills trees.

Cut-offs

Cut-offs are barriers or dips constructed across paths on slopes to divert any water flowing down the path to run off at the side. They are also known variously as diversion drains, waterbreaks, jarnocks, by-sets, grips, waterstops, and water-bars. In America they are known by rather more colourful terms, including bleeders, kick-outs and thank you ma'ams.

Their construction may be similar to that of a cross drain, which takes water from one side of the path to another. A cross drain may also act as a cut-off if it is on a sloping path.

The three basic ways of making a cut-off are as a drain, a ridge or a barrier with cut and fill. Drains and ridges are used on hard surfaced paths.

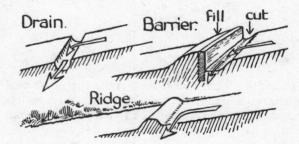

Position

Look carefully along the edge of the path for fans or scatterings of pebbles or silt, which indicate where water flows off during high rainfall. The path immediately above will probably show signs of gullying, where the material has been washed away. Install a cut-off at the point where water already flows off, and some more at intervals of 3m to 5m up to just above the point where gully-ing begins.

Choose a position where there is already a slight dip, or where there is a boulder or tree root against which a barrier can be wedged. Make sure there are no obstructions off the edge of the path which will block the flow of water, but beware of directing the water off a steep edge of unconsol-idated material, as it will quickly erode.

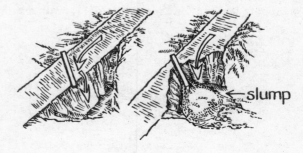

A cut-off should be installed just above the angle of a hairpin bend. Take care that the outflow does not look like a gullied path, or walkers will be tempted to make short cuts. A short, deep trench or a length of buried plastic pipe should take the water off without encouraging people to follow.

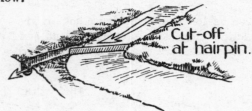

Angle

The angle should be set so that the cut-off is self-cleaning, and neither erodes nor fills with debris. An angle of 30 to 45 degrees is suitable for most situations.

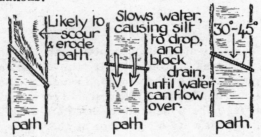

Height

The bar must be high enough to divert the flow, but not so high that walkers regard it as a barrier. The cut-off should not be visible from below, but should merge into the profile of the path.

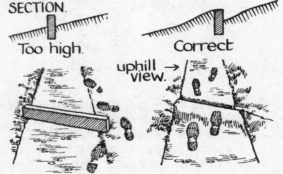

In the Peak District, where many cut-offs are made of embedded stones, it has been found that by dipping the bar towards the middle, walkers

are encouraged to step over at the middle and not to walk around the ends of the cut-off.

Length

Extend the cut-off at least 300mm on either side of the path, to discourage walkers from going around the ends. The cut-off will need extending further if the path widens with use. The lower end of the cut-off should lead into a ditch to take the water quickly down the slope. The ditch should be about 300mm wide with as steep sides as possible, and at least a metre long.

CUT - OFF DRAINS

Cut-off drains can be made out of wood, slate or stone. The method of construction is the same as for a cross drain, and is described on page 68. Pre-cast concrete and clay channels are also available. The main advantage of cut-off drains is that they are flush with the path surface and so present no barrier to the walker. Concrete cut-offs should be used on paths that are designed for wheelchairs and prams, as their narrow opening prevents the wheels getting stuck (p182 7.6). Concrete cut-offs need clearing out frequently.

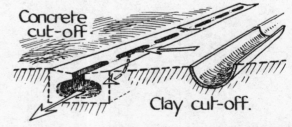

In an emergency, when water is flooding down an unconsolidated path, it is worth using any tool available, spade, stick or boot, to dig out a rough drain and try to divert the water off the path. This should be replaced as soon as possible by a properly constructed cut-off.

CUT - OFF RIDGES

This technique is used on tarmac or concrete paths, but is not usually appropriate on rural paths. Ridges have been used on stone-pitched paths in the Peak District, and although they successfully divert the water, they tend to merge rather too well with the surfacing and could easily be stumbled over. A stone drain is probably the best solution on a pitched path.

Ridge in pitched path.

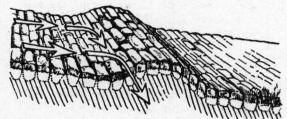

Ridges of earth and stone are not recommended as they are too easily washed away or trampled flat.

CUT - OFF BARS

These are the type of cut-off commonly built on trails in America, where timber is abundant.

1. Peel the log with a peeling spade or light axe, as the bark increases the speed of timber decay if left in place. Stakes can be made of timber up to about 75mm diameter, cut into 450mm lengths.

2. Dig a trench to a depth of half the diameter of the log, and a width of double its diameter.

3. Notch the uphill side of the log to accept the stakes so they will not clog the cut-off.

4. Seat the log securely in the trench, and drive the stakes in at an angle to hold the log. Saw off flush any extra that cannot be driven in.

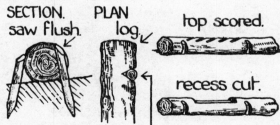

SECTION. saw flush. PLAN log. top scored. recess cut.

Notch on uphill side.

5. Score or recess the top of the log to give a rough surface that boots will grip.

6. Pack excavated soil and rock on the lower side of the log. Boulders placed at either side will give extra strength and discourage walkers from going around the ends. Seat the boulders in carefully so they appear 'natural'. Dig the outflow ditch.

SECTION.

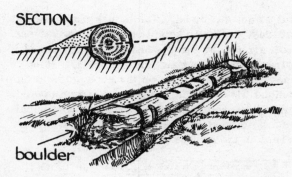

boulder

Sawn timber

Sleepers or other heavy planks (175mm x 50mm minimum) can be used in a similar way to logs. Use stakes of timber 50mm x 50mm, knocked in on the lower side only of the plank. Weather the tops of the stakes, and nail them to the plank. Use square or half-round stakes in preference to round, as they give a stronger joint.

Section Stake→

Sleepers can be nailed, but are better secured by angle iron or metal pins. The strongest anchorage is made by drilling the sleeper with an auger and bit, and knocking the metal pins through it. Alternatively, the timber can be drilled part-way, and then positioned over the accurately secured pins. If using angle iron, nail it to the lower corners of the plank with 50mm galvanised clout nails.

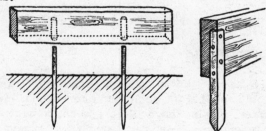

Slate

In some part of Snowdonia old fence slates are available, and make excellent cut-offs. They are usually set in rocky ground where they can be securely wedged without the need for stakes. If staking is required, use metal pins.

Stone

These are made of a double row of slab-shaped stones with overlapping joints. Take time to seek out the best available stone, preferably of about 40mm thickness. The stones should be set in at least $\frac{2}{3}$ of their height to prevent movement. In soft ground lay large stones in the bottom of the trench to prevent the slabs sinking. 'Heel' stones can also be put in on the upper side to stop the slabs tilting. Compact the excavated material against the lower side of the bar. As noted above, a slightly dipped profile will encourage walkers to cross in the middle of the bar.

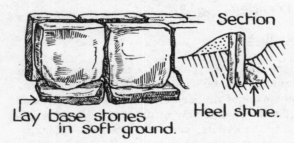

Section

Lay base stones in soft ground.

Heel stone.

Maintenance

It is difficult to construct a cut-off at the exact angle to be self-cleaning, and most will need cleaning out several times a year. Scrape away the debris on the upper side with a mattock or spade, and pack it against the lower side of the bar. Clean out the drainage ditch off the end of the cut-off. If the ground above the bar is gullying, more cut-offs should be built higher up the path. Do not deepen the trench below the bottom of the bar, or it may be undermined.

SECTION.

Cross Drains

These take the flow of water across a path. They are only suitable for small and intermittent flows of water, or else they will have to be too wide to step over easily. Any permanent flow, except from a nearby spring, will be too big to carry in a cross drain, and must be culverted or bridged.

Size

The following sizes are suggested to cope with various conditions on a path:

a   Small. Drains with a diameter of about 100mm, usually concrete or clay channels. These are only sufficient to take small seepages of water, that cause muddy patches after heavy rain. These are likely to be in lowland and low rainfall areas with porous soils.

b   Medium. Drains with a cross-section of about 100mm by 100mm, usually of wood or stone. These should be suitable for places where water flows across the path during heavy rainfall but not in sufficient volume to wash away the path.

c   Large. Drains with a cross-section of over 300mm by 300mm, usually of stone. These must be constructed at places where a torrent of water crosses the path after rainfall or snow-melt, probably damaging the path. These should be built as big and strong as possible, while still being negotiable by the walker. Fortunately, these sort of flows are typical of mountainous and rocky areas where the stone to build them is available.

A wide and deep cross drain will also serve as a barrier, which may or may not be desirable. On a path designed to be used by elderly, disabled or those with young children, culverts will have to be built.

Position

If possible, walk the path in wet weather and mark the points where water is crossing or seeping onto the path. Otherwise, inspect the edges and surfacing carefully for signs of where water has crossed.

Usually the drain will take the shortest route, straight across the path. If the natural flow is at an angle, ensure that it cannot divert and flow down the path. Dig the drain extra deep, and reinforce the lower edge with boulders.

Stream may divert.

Reinforce with stone.

Wooden cross drains

These can be made in a workshop and then taken to the site to be installed. Pressure-treated timber should be used, as the timber has to withstand alternate wetting and drying, and contact with the soil. As well as making the drain stronger and less likely to collapse due to scour, the solid bottom makes it easier to clean out. If timber is in short supply, spacers can be used at the bottom instead. The spacers at the top are the weakest part of the construction, so use good pieces of wood, without knots or faults, and install with the grain as shown. If one does break in use, another can be nailed across the top. The advantage of these drains is that they can be quickly installed without major excavation. They should be set flush with the surface of the path, so they do not cause any hazard to walkers.

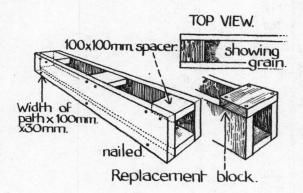

TOP VIEW.

100x100mm spacer.

showing grain.

Width of path x 100mm x 30mm.

nailed.

Replacement block.

Stone cross drains

These make very strong and permanent drains if sufficiently heavy stones are used. The stones should be big enough to form both the side of the drain and the surface of the path, and heavy enough that they do not move when trodden on. Thin stones that merely line the drain are not sufficient and will soon get displaced.

The drain should be at least 200mm wide, and as deep as the stones allow. As with all work involving large stones, it is worth taking trouble to ensure the stones are perfectly seated so they will not move.

To protect against scour, the drain can be built by first lining the bottom with large slabs or boulders which are then held firmly in place by the side stones.

backfill with stones

As well as taking torrential flows of water, stone drains can also be built to take seepage water from damp flushes on mountainsides. The stones placed at the inlet help keep the water table high in the flush during summer, and prevent material washing in and blocking the drain.

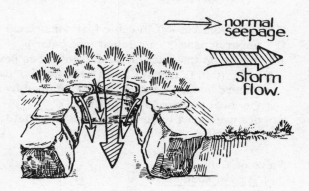

normal seepage.

storm flow.

Cross drains that drop steeply from the path should have a boulder with a slab, called a splash plate, to protect the drain from scour.

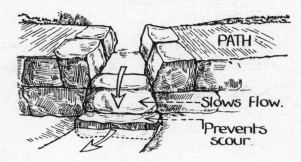

PATH

Slows Flow.

Prevents scour.

Culverts

These take water under the path, and are built where the flow is too great for a cross drain, or where vehicular access is needed. They do need regular maintenance to keep them clear, and depending on the situation and the materials available, it may be more effective to construct a simple bridge.

Size

Culverts block easily, and should always be built as big as is practical. It is possible to estimate the diameter required for given run-offs or areas, using the formula and tables below. It is stressed that these are only estimations, and it is impossible to predict exact flows. The best advice on choosing the size of pipe is to compare results from the two different tables below, investigate any similar pipes near to the site, and make an informed guess! If nothing else, the tables below give an indication that culverts must always be bigger than you expect.

The tables are from the Design Manual issued by Armco Ltd (p182 7.3). Imperial units are given in the absence of simple metric equivalents.

$$Q = CiA \text{ where}$$

Q = maximum run-off in cubic feet per second
C = imperviousness coefficient (see below)
i = maximum intensity of rainfall in inches per hour (this is usually taken as being 1.5" but a more accurate figure can be obtained from the local meteorological office)
A = drainage area in acres (from map or field estimation)

Value of C

Type of surface	C
Rocky	0.9
Rocky on ½ the area	0.7
Impervious soil	0.65
Slightly porous soil	0.4
Moderately porous soil	0.2
Wooded areas	0.2

Example: for an area of 5 acres of partly rocky ground;

$$Q = 0.7 \times 1.5 \times 5$$
$$= 5.25 \text{ cubic feet per second (cusec)}$$

TABLE 7b

Diameter of culvert in inches/mm	Discharge in cubic feet per second	Height of flow in culvert
42 — 1050	60, 50, 40	1.0
	30	
36 — 900		0.9
33 — 840	20	0.8
30 — 750		
27 — 675	10	0.7
	8	
24 — 600	6	
	5	
21 — 525	4	0.6
	3	
18 — 450	2	
15 — 375		0.5
	1	
12 — 300		
9 — 230		

Using the value of Q, look up the diameter of culvert required in table 7b.

To use the table, place a ruler through the measured discharge in cubic feet per second (Q) and the acceptable maximum height of flow in the culvert, and then read off the culvert diameter from the left hand column. For example, the measured run-off of 5.25 cusecs would required a culvert of 18" diameter, flowing full.

An alternative estimation can be made by using table 7c below.

TABLE 7c TALBOT'S FORMULA

Diameter of culvert		Area of waterway opening in sq ft	ACRES DRAINED		
			Mountainous country	Rolling country	Level country
12"	300mm	.79	¾	3	6
15"	375mm	1.23	1	6	11
18"	450mm	1.77	2	9	18
21"	525mm	2.40	3	14	28
24"	600mm	3.14	5	20	39
30"	750mm	4.91	8	36	71
36"	900mm	7.07	14	59	115
42"	1050mm	9.62	20	89	175

Tables 7b and 7c give comparable readings for rolling and level country (wooded and porous soils), but table 7c gives a larger estimate for mountainous (rocky) country. Build to the larger size if possible.

Headwall

A headwall should be constructed at either end of the pipe to protect it, and to stop any flow through the fill around the pipe which will cause erosion. If the pipe is over-topped, a strong headwall will act as a dam and cause increased flow through the pipe due to greater water pressure.

Usually the height of the headwall should equal the diameter of the pipe plus 300mm, to take it level with the surface of the path. Headwalls can be constructed of stone or brick.

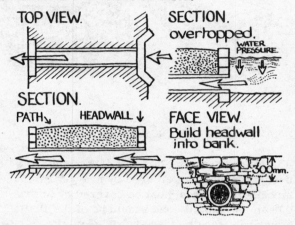

If no material is available to build the headwall, the bank should be sloped as shown below. A headwall is preferable however, as a sloping bank will require a greater amount of fill and a longer and more expensive pipe.

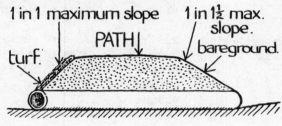

Apron

If the headwall is being mortared, it is worth constructing aprons. Aprons are usually specified for pipes over 440mm diameter.

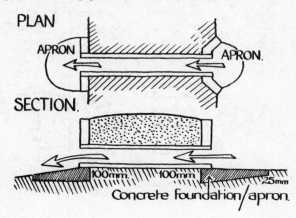

Vehicular access

If there is any chance of a vehicle crossing a culvert, even if it is not legally entitled to do so, then build the culvert strong enough to withstand it. A piped culvert can be broken by the passage of a single vehicle, and is awkward to repair. If it is possible to get a vehicle up the path, it is likely to be useful to do so for future path management, so put in a pipe long enough to give a path width of 2½ metres. The pipes listed for culverts on page 63 are strong enough for normal vehicle weights if protected with at least 300mm of backfill. The Ministry of Agriculture specification for a culvert for heavy farm machinery is given on page 72.

PIPED CULVERTS

Except for the seepage culverts described below, pipes should be at least 225mm diameter. Use either clay, concrete or steel pipes. The choice will depend on local availability, and on the means of transport to the site.

1   Excavate trench if necessary. Existing stony stream beds and stony subsoils should provide sufficient support. In peaty soil the pipe should be supported by a layer of 10mm -15mm diameter aggregate.

2   Lay concrete foundation/apron if specified. Lay pipe and build up headwalls around and above the pipe.

3   Compact stone-free filling (see p64) at the sides and top of the pipe. It should be put down in layers not more than 150mm thick and thoroughly consolidated, to a total depth over the pipe of at least 300mm.

4   Surface with sub-base, base and surfacing, crowning each layer so that water runs off.

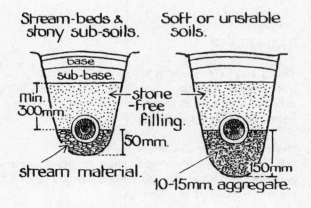

Seepage culverts

These are very simple culverts, designed to take seepage flows from small bogs and mountain flushes. These often occur frequently along a path, and a cheap, simple solution is needed that will dry out the path without draining the bog or marring the appearance of the area. This is the only situation where pipes below 225mm diameter are recommended for culverts. 100mm diameter uPVC pipes are suitable, being light and easy to install without major excavation.

← cover with at least 100mm of material, dug from nearby, if necessary.

STONE CULVERTS

These are traditional in many upland areas of Britain where suitable stone can be found. The edges of the culvert should be constructed in the same way as a cross drain, using the largest boulders which can be found and moved into position. Make the inside edge as smooth as possible, so that debris will wash through and not wedge inside. Stone culverts sometimes also have a solid stone bottom to prevent scour.

PATH

STREAM.

BOX CULVERTS

These can be made of wood, slate or stone. Untreated elm is good for culverts that take fairly constant flows as it is durable in water. Otherwise, use treated softwood (see p183).

Wooden culverts can be nailed together, in a similar way to the wooden cross drains described on page 68. They are not recommended for culverts crossed by vehicles.

Stone and slate culverts rely for their strength on

being carefully constructed and backfilled. Align all the stones so there are no protruding edges which will catch debris. Pack angular stones along the sides and up to the top of the culvert, and then cover with at least 300mm of stone-free material, well tamped down.

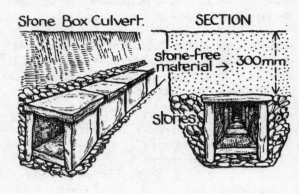

Stone Box Culvert. SECTION

stone-free material → 300mm

stones

HEAVY VEHICLE CROSSINGS

The following is the Ministry of Agriculture's design for culverts crossed by heavy farm machinery. Pipes must be covered by at least 900mm of backfill, and should be of the British Standard strength class shown below.

	Diameter	BS strength class
Clay pipes to	225mm	standard strength
BS 65/540	300mm	extra strength
	375mm	standard strength
Concrete pipes	225mm	standard strength
to BS 556	300mm	extra strength
	375mm	" "
	450mm	" "
	525mm	" "
	600mm	" "

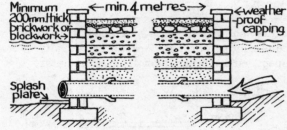

Minimum 200mm thick brickwork or blockwork

← min. 4 metres →

weather proof capping

Splash plate

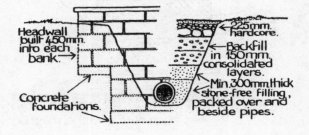

Headwall built 450mm into each bank.

Concrete foundations.

225mm hardcore.

Backfill in 150mm consolidated layers.

Min. 300mm thick stone-free filling, packed over and beside pipes.

Side Drains

These are typically constructed along the side of contour paths in upland areas, or along sunken paths where water cannot easily be drained away from the path. They intercept surface and sub-surface flow and carry it away downhill.

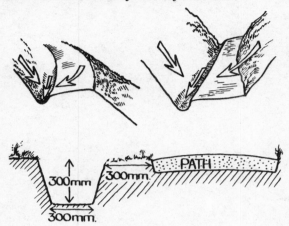

They may also be constructed by paths that run directly down slopes, and which may otherwise themselves become watercourses.

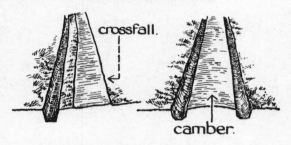

Material from the ditch can be used to build up the surface of the path, or scattered below the path. Re-use any turf on eroded parts of the path.

Dig the ditch as close as possible to the edge, whilst ensuring that the side of the path will not collapse into the ditch. To prevent this happening, side drains can either be lined or filled.

Lined drains

The design below is used beside a pitched path at Beacon Fell, Lancashire.

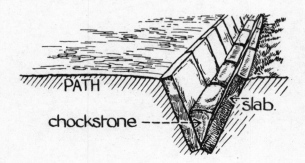

A wider drain can be made with either stone slabs or mortar and stone.

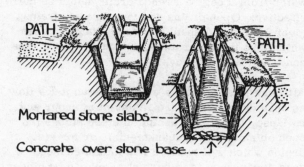

Alternatively, the bank can be protected from scour and collapse by stone pitching. This method has been used at Castle Eden Dene, Durham.

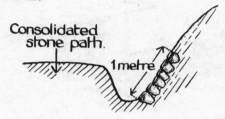

French drains

These are ditches filled with permeable backfill of clean gravel, stones, stone chippings, slag or clinker, graded from diameter 5mm to 50mm. It is important that the material contains no particles smaller than 5mm, or it will rapidly block.

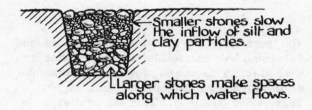

In peaty soils, which do not contain the fine clay and silt particles that can wash in and block the drain, larger single-size aggregate up to 100mm diameter can be used.

When deciding on open or French construction, consider how the side drain works. If it is a carrier for water from a spring, or if it intercepts

surface flow, an open ditch is recommended. If
it intercepts sub-surface flow, or carries seepage
water from a bog, a French drain should be more
effective. Open drains and long French drains
(over 10 metres) should lead into watercourses
which take the water away. Short French drains
can lead into soakaways.

Although French drains cannot take as much flow
as open drains, they are resistant to scour and
collapse, and so are useful on steep slopes (as
seen on motorway embankments), or for side
drains which will receive little maintenance. If
good appearance is important, they can be hidden
by covering with turf, but this lessens their
ability to intercept surface flows.

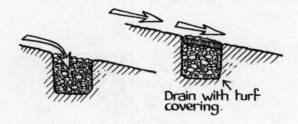

Drain with turf
covering.

The effectiveness of French drains can be
greatly increased by laying a permeable uPVC
pipe inside the drain.

permeable
plastic pipe.

The 'permeable fill' of French drains does tend
to become less permeable as fine particles are
washed in. This can be prevented by lining the
drain with a permeable membrane such as
Terram 500 (see p78). This can be cost
effective as it allows a smaller amount of permeable fill to be used to make the same capacity
drain, and lengthens the life of the drain. Also,
coarser ungraded fill can be used, as it does
not need to act as a filter.

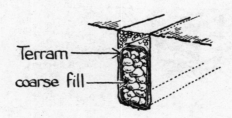

Terram

coarse fill

8 Surfacing

Most paths have developed on surfaces that are resistant to a certain amount of trampling. Examples are paths on grassland, beaten earth paths through woodland and mountain paths on rock. Where unavoidable, paths cross peat bogs, marshes or sand dunes which are unstable and easily damaged by trampling. The addition or replacement of surfacing material becomes necessary either when the natural surface is damaged or destroyed by more use than its natural structure can bear, or where an unstable material needs protecting and strengthening.

The word 'surfacing' is usually applied to any material laid down as a path, but in the following chapter the parts of a path are termed as shown:

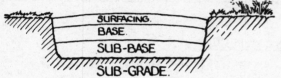

Whether or not the three different types of material are needed will depend on the natural 'sub-grade', the drainage, the expected use, and the material being used to make the path. In many cases only a single layer will be necessary. A sub-base will be needed in unstable or wet ground such as peat, or on paths likely to flood which must be raised above flood level. In most situations the bedrock, subsoil or topsoil will be strong enough to support the base and surfacing. The base is the main load-bearing part of the path, and in most jobs will comprise the bulk of the material used.

The surfacing provides a surface which should be comfortable to walk on, non-slip and resistant to trampling. It may be unsealed material such as gravel, sealed such as tarmac, or solid such as stone pitching.

The majority of paths do not need surfacing, and would be spoilt by any attempt to do so. However, a small proportion of paths, mostly the very popular ones, are being damaged by the lack of a sound surface. Surfacing is usually necessary for paths designed for heavy use in recreation areas and country parks.

The Path Profile

It is important that new paths should built with a cross-fall or camber to shed water off the path.

On existing paths that are badly drained, the construction of a camber or cross-fall is the simplest method of improving drainage, and one which can be done to the natural path profile whether or not surfacing material is being used in addition. In most situations this will be done by hand using spades and shovels, but on wide paths and for major improvement schemes a tractor with a grader (see p181 6.2) can be used. During the Brecon Beacons Pony Trekking Project (see Bryant, 1978), this technique was found to be vital for the success of any subsequent surfacing work. Surfacing material simply laid on a poorly drained path will not make any long term improvement, as it will get trodden in and dispersed under waterlogged conditions.

The same profile of cross-fall or camber should be followed through any layers of surfacing.

Cross-fall

Make the cross-fall about 1 in 30. For example, on a 2 metre wide path, level using a block of wood 60mm high.

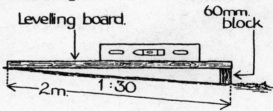

On a contour path, the cross-fall can either shed water straight down the slope, or into a side drain leading to a culvert. The choice depends on the general conditions listed below.

'Lowland'.
Porous soils, with surface run-off in streams. Removes flow from path only.

'Upland'.
Impervious soils, with surface run-off not in streams. Takes run-off from path and slope.

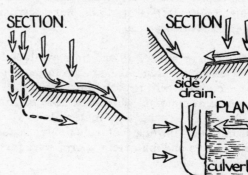

Camber

The best use of a camber is on causeway-type paths on flat marshy ground, where it is necessary to build the path up, and to provide as much drainage as possible. A camber can be made using material from ditches dug on either side. Cambers are also usually made on forest tracks and rides, which are wide enough to make a functional camber that is comfortable to walk or drive on. Most footpaths are relatively narrow, and a camber steep enough to work is awkward to walk on.

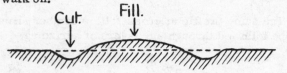

The camber described above is made by cutting and filling the ground itself. Cambers can also be made of the surfacing material, but these tend not to be very permanent if of loose material such as gravel or crushed stone, as the camber quickly gets trampled flat. Hard surfacing such as stone pitching can be constructed with a camber.

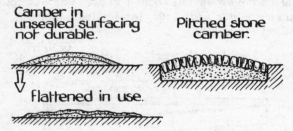

Construction and Edging

When choosing the method of building the path, consider the following:

a The physical properties of the land. This includes the strength and permeability of the soil, the gradient of the land, and the surface run-off. See Chapter 3.

b The appearance of the path. Hard edged or raised paths are likely to be more obtrusive than excavated paths.

c Machinery, materials and time available. For example, a small increase in the depth of excavation will greatly increase the volume of material which will need shifting (see p86). Transport of materials is usually the limiting factor, so only choose techniques and materials appropriate for the site.

The following diagrams show different methods of building a path.

Excavated

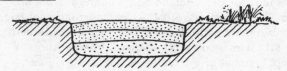

This will be necessary for heavily used paths on poorly drained, thin or fragile soils. Examples include heavy clays, thin soils over chalk, and peat and sand. The line of the path can be cut by hand or using an excavator. The 'cut' must be used elsewhere as 'fill' along the path, or in hollows nearby. The usefulness of the fill is an important factor in deciding whether or not to excavate.

Edged and filled

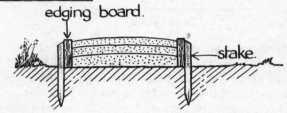

This method can be used to raise a path above a high water table. The raised edge does provide a slight psychological barrier which discourages walkers from stepping off the path. No excavation is required, except to provide firm footings for the edging.

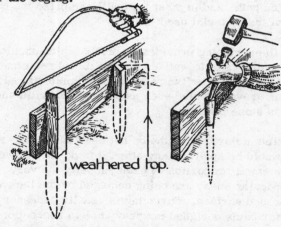

The edging can be of sawn timber, sleepers, logs or boulders. Sawn timber about 100mm x 25mm, treated with preservative (see p184) is suitable. Secure with square wooden stakes, angle iron or metal pins. Knock in a few test stakes to determine the length required to give a secure hold. Weather the tops of wooden stakes, and knock angle iron or pins below the top of the edging, using a punch if necessary. This allows for any settlement, which could expose the tops of the pins.

Log edgings have the advantage of giving a more 'natural' edge, and can be fixed unobtrusively (see p81). They may also provide a cheap solution for situations where the edging is only designed to be temporary, until the path has been compacted with use. In this case the logs can be laid on the ground without staking, and taken up for re-use after a year or so. This is only suitable for situations requiring a moderately strong path, such as woods and dry heaths, as the surfacing material will not be very thick.

Boulders are not recommended unless they can be effectively hidden by vegetation. Unfortunately, they are usually the easiest material to use on open hillsides, where they give a result more appropriate to a seaside garden. Even if initially well covered with the base and surfacing material, they tend to get exposed as the material is compacted or eroded, leaving a line of often conspicuous white boulders. If surfacing is necessary on open hillsides, an excavated or informally edged path is recommended.

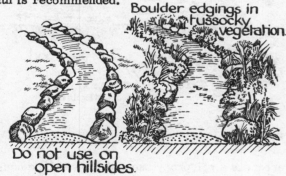

Boulder edgings in tussocky vegetation.

Do not use on open hillsides.

<u>Informal edge</u>

This is suitable for material such as hoggin or chalk which sets solid when compacted. The path will be weak at the edges, and tempting to walk off, so this is only appropriate for fairly resistant situations such as grassland, dry heath and wood.

<u>Paths along contours</u>

The usual technique on steep hillside is to cut,

fill and revet. Revetments are described in Chapter 11.

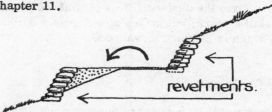

revetments.

On shallow slopes, the path can be edged along the lower side, using any of the edgings described above.

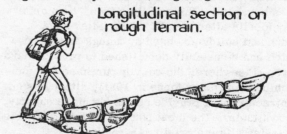

Sub-base Materials

Sub-base material is used on very rough or wet terrain, to even out the ground, or provide a solid bottom for a path over soft ground. In most cases a base and surfacing layer is required on top.

BOULDERS

Where available nearby, these can be used to build or improve contour paths, and to even out the gradient of paths crossing rough terrain.

Longitudinal section on rough terrain.

Parts of the Watkin Path to Snowdon were repaired during 1981 using the method shown below, with boulders up to 4m in length being positioned by

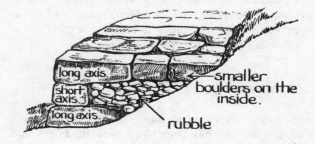

long axis.
short axis.
long axis.
smaller boulders on the inside.
rubble

Tirfor winch. Flat-faced boulders were also used to form the surface. This is time-consuming and skilled work, but makes an extremely durable path which is pleasant to walk on.

MORTARED STONE

Mortaring will be necessary to strengthen a stone sub-base on very exposed contour paths, or by streams and rivers. Use a sand:cement mix of 4:1, and use sharp sand on the outside as its colour is less obtrusive than most builder's sand.

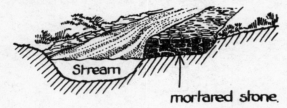

STONES IN PEAT

Popular paths on deep and wet peat are amongst the most problematical. Stone sub-bases have been attempted, but have not been successful on their own as the stone simply sinks, and the path becomes wet and rough and is consequently avoided. Paths on deep peat that are heavily used need a membrane of Terram or similar, or a floating mattress of brushwood (see below).

HARDCORE

This is usually building rubble, and provides a good sub-base material as it is strong and free draining. It should be used with caution on acidic sites because leaching of lime from the mortar may locally alter the pH. Some sites have redundant buildings, hard standings and so on, which can be usefully demolished to provide hardcore. Elsewhere, the cost of purchase may be prohibitive (£6 per tonne in 1981), although some contractors will provide odd loads more cheaply, as available. The local council may also be able to assist with material from road works.

Hardcore should always be used with care in the countryside, as it is an unsightly material. It is probably best used in excavated paths, where it can be covered with an adequate layer of base and surfacing, and there is no chance of it becoming exposed by use.

BRUSHWOOD

This is a revival of an ancient road-building technique, and has been used recently to try and stem the damage of moorland on parts of the Pennine Way, near the Snake Pass. Trampling had destroyed the vegetation and reduced the peat to a structureless sludge, so pine brashings and bundles of heather were laid down to form a 'mattress'. A lot of material was needed initially, and more must be added until it consolidates to form a dry causeway. It does not make a very comfortable walking surface and looks untidy, but most walkers would appreciate that it does at least contain the damage, and is better than knee-deep sludge as a walking surface. It could be improved by covering with a base and surfacing material, although none has yet been tried. It may be possible to use straw or bracken, either as a sub-base, or as a base over brushwood. Bracken could usefully be cut from areas where it is invading dry heathland.

Old chestnut paling fence has been used successfully in a similar way on deep wet peat. This is quicker to lay than brushwood bundles, and gives a neater and less hazardous walking surface. On very wet sections it may need to be supported by brushwood.

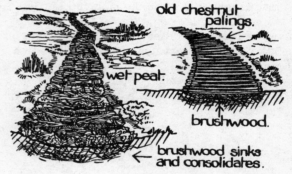

TERRAM

Terram is a polypropylene and polyethylene non-woven permeable membrane, manufactured by ICI for various uses in civil engineering (see p182 8.1). It is available in several gauges, of which 500 and 700 are considered here. Terram has been used by many different authorities for the construction of footpaths and bridlepaths. Terram acts as a sub-base, and prevents the base layer from sinking or being trodden in to wet soil or peat.

Gauge	Thickness	Width	Length	Price/roll (1981)
500	0.5mm	4.5m	200m	£250.00 approx.
700	0.6mm	4.5m	200m	£220.00 approx.

To lay Terram:

1   Clear the line of the path. Dig up or treat chemically any woody stumps (see p58)

and cut through tussocky grasses, rather than pull them out by the roots. Pulling them out will leave holes which must be filled and levelled.

2   Cut the Terram to a width 50mm wider than the finished surface. This is easiest to do by cutting through the roll with a hacksaw. Terram can also be cut with snips or heavy scissors, but this is hard work.

3   With a sharp spade, cut along the edge of the path.

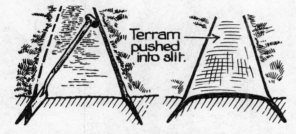

4   Unroll the Terram, tucking the edges under the turf. This hides the edge, and holds the Terram in place while the base and surfacing material are being laid.

5   Alternatively, if side drains are being dug, the turf can be used to anchor and hide the edge of the Terram. The use of boulders to anchor the edges is not recommended, as this always gives an inappropriate 'garden path' appearance.

WYRETEX

Wyretex is made of polypropylene and galvanised wire twisted together in strands, and woven into a mesh. It is available in several different grades (p182 8.2), of which grades 8 and 9 are considered here.

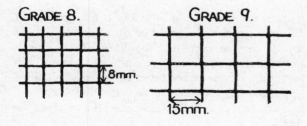

Grade	Width	Length	Price/roll (1981)
8	1.5m	100m	£180.00 approx.
9	1.5m	100m	£127.50 approx.

Three different construction techniques using Wyretex are shown below.

Excavated

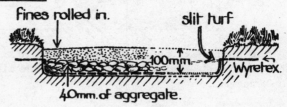

Part excavated

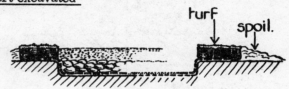

Pinned

This method has been successfully used at Risley Moss, near Warrington, to make paths through damp, acid woodland.

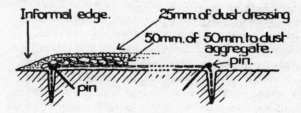

Unsealed Surfaces

The materials described below can be used for the base or surfacing course. They are laid either on a constructed sub-base, a membrane, or direct onto the ground if suitable.

The materials act in one of the three following ways:

a   'Soft' stone, such as chalk or limestone, which shatters when rolled to form an almost solid surface.

b   Stone with clay or minute particles called 'fines', which bind when rolled and set to form an almost solid surface. Examples are hoggin and self-setting gravel.

c   Material without fines should be graded and laid in layers. A useful rule of thumb is that the size of the material should be half

79

that of the available depth.

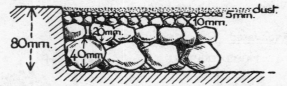

If the material is simply thrown in together, a structurally weak and uneven surface will result.

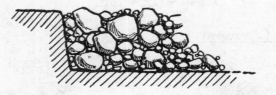

LOCAL MATERIALS

This describes materials which are obtained on site or nearby. Any material which acts in either of the three ways listed above may be suitable.

Streams

Material from 'young' or upland streams is often suitable as the stones are still angular and have not yet been rounded by the action of water. Shovel material from the edge of the water or from shallow pools into buckets, and lay it without grading. The fine sediments should bind the angular stones as the material is packed down. Old buckets with holes are useful for this job as they save carrying water unnecessarily. Beware of taking so much from one place that the course of the stream is altered.

Outwash deposits

Outwash deposits are an excellent source of material, and are usually conveniently graded during deposition. Carefully shovel up and lay the material in layers, starting with the coarsest and finishing with the finest.

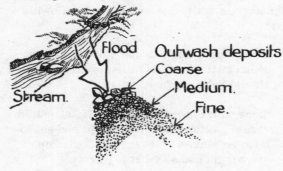

Scree

Scree material is very angular, and is suitable

for the base course. It is usually washed clean of any fine sediments, and must therefore be graded and laid in layers.

Screens for grading surfacing material can be made out of galvanised Weldmesh or similar, nailed to a wooden frame. Revolving drum screens can also be used.

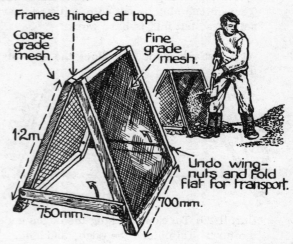

Borrow pits

These are small pits dug nearby in suitable material. Many upland paths and tracks were built by this method, which has the great advantage that material only has to be carried a short way.

COMMERCIAL SUPPLIES

Crushed stone

The type depends on what is locally available, and names and suitability for paths vary enormously. The usual specification for base material is 60mm to dust, which is laid in a layer about 100mm deep and rolled. A surfacing layer of 'fines', 'dust' or 'blindings' (fine crushed stone) is then laid about 25mm thick.

Inspect the material if possible before purchase. The base material should be of a mixed size, and both it and the surfacing should be of a colour which blends with the surroundings where it will be used. Special care must be taken when using limestone on acidic soils as water leaching through it may locally raise the pH and alter the flora.

Quarry bottoms

This is the waste material from the quarry bottom and is the cheapest type of quarried stone. It is obviously very variable, but should contain a lot of fine sediment and will bind without grading.

Hoggin and self-setting gravels

These are gravels with a high clay content that bind when rolled to form a very hard surface. If obtained locally they should blend reasonably well with the surroundings, but some hoggin does have an orange colour which may not be suitable. These gravels are often used for both base and surfacing, and for best results should be rolled several times with a vibrator roller.

A total depth of 100mm for base plus surfacing should be sufficient for footpaths, and would normally be laid directly onto the soil or subsoil if well drained. Poorly drained areas or bridlepaths will need a sub-base of hard core or stone. Hoggin is not permeable, and should be laid with a camber or cross-fall. A final dressing of pea gravel rolled in will protect the surface against scuffing.

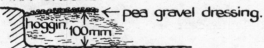

FIRM GROUND.

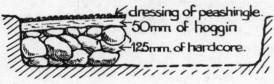

SOFT GROUND or BRIDLEWAYS.

Stone as-dug

This usually contains clay or subsoil. Always inspect before purchase if possible.

Wood chippings

These may be bought from sawmills as a by-product, or can be made from local timber by using a wood-chipping machine.

Chippings are most suitable for sheltered woodland paths. A likely site would be a heavily shaded path where grass cannot grow, and which becomes muddy with heavy use. The layer of wood chippings prevents churning of the earth, and lifts the path above badly drained soil. It also provides a 'waymark' through dense woodland, which many people find intimidating to enter.

Wood chippings give a pleasant springy surface, which is especially attractive to horse-riders. Unfortunately, the horses rapidly churn the surface, so chippings should only be used for bridlepaths if frequent maintenance can be done, and where separate provision is made for walkers.

The following construction has been used at Beacon Fell Country Park, near Preston. It has withstood fairly heavy use by walkers for over three years, without maintenance. The path follows the contours of a gently sloping hillside in coniferous woodland.

1 Culvert all water that crosses the path, and ensure that none can flow down the path, as the chippings are easily washed away.

2 Place and stake the log edgings as necessary on the lower side of slopes.

3 Spread the chippings to a maximum depth of 150mm. They do not need rolling or compacting.

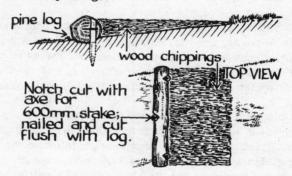

Woodland Mulch is a product made of bark and wood chippings which is composted for several months under high temperatures (p182 8.3). This gives a durable coarse-textured material, light in colour and of neutral pH. It can be used for paths in a similar way to wood chippings.

Non-composted bark is not suitable for surfacing as it becomes very slippery when wet.

INDUSTRIAL WASTE PRODUCTS

Basic slag

This is a waste product from steel works, and makes a good base or surfacing material. It is usually available in 20mm to dust, and should be laid and rolled over a firm sub-base.

Cinders (clinker)

This can be obtained from coal-fired industrial premises and is suitable for the base course.

RAILWAY BALLAST

Many old railway lines are now being opened as footpaths and bridleways. In some cases the ballast has been removed, leaving an artificial sub-base which is usually free draining and should support a good grass cover if there is sufficient light. If the ballast is still in place, it is likely to be loose and uncomfortable to walk on, and needs a surfacing of clinker or crushed stone of about 20mm to dust well rolled in.

STONE TO AVOID

Pea gravel, washed gravel and road chippings are not suitable for either base or surfacing. Pea gravel and chippings are too small and angular, and washed gravel too rounded, to form a strong load-bearing base. If used as surfacing the material moves and 'crunches' unpleasantly underfoot, and the path will invariably be avoided.

If you have to deal with an unsuccessful path made in this way, the best solution is to scrape up the chippings and stockpile them nearby. Make a new base of hoggin or crushed stone, and use the chippings to dress the surface from time to time.

Chippings or pea gravel may have use just to dress a beaten earth path to reduce formation of mud, but this is not a permanent solution to the problem. Only put down enough to just cover the surface so it gets trodden in. This will have to be maintained frequently.

MAINTENANCE

Most unsealed surfaces will need annual maintenance.

a   Note any patches which are often water-logged, or places where surfacing or base material is washed away. Install culverts or cut-off drains as necessary.

b   Replace surfacing or surface dressing as necessary, as it will get worn away with use even on perfectly drained paths.

c   Replace damaged edgings if they are still required to contain the path material.

d   Presuming that surfacing has been necess-itated by use, trampling should keep down weedy growth on the path. Watch out though for thistles, which are very persistent and can come up through a substantial depth of stony material. Treat with a herbicide such as glyphosate.

Sealed Surfaces

Tarmac, concrete and asphalt are the hardest-wearing surfaces for footpaths, and may be needed on heavily used urban paths, and some paths in country parks and other recreation areas. Sealed materials can be used on slopes which are too steep to hold unsealed materials, but not steep enough for steps. Tarmac and asphalt are not materials which are likely to be used by volunteer parties, so their use is not explained here. Concrete is an easier material to use than tarmac or asphalt, and its colour is usually more sympathetic to a rural situation.

CONCRETE

Sections of path which cannot be drained or which are subject to stream or river erosion can be concreted. See Appendix D for suitable mixes.

Formwork paths

This is the standard method of making concrete paths, and results in a hard-wearing but rather unattractive path with straight edges. It will mellow with time if the edges are allowed to grow over. Always sink the path so the top of the concrete is level with the ground surface.

1   Remove soil from the line of the path, allowing at least 75mm extra on either side for the formwork. Firm ground can be rolled and then the concrete laid direct. Soft ground must be excavated further to give sufficient depth for a 50mm sub-base of hard-core or stone. The concrete should be at least 75mm thick.

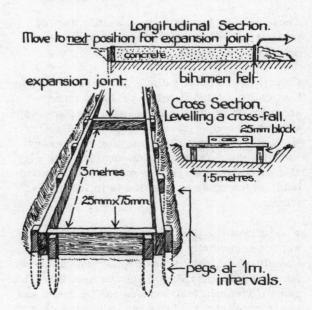

Longitudinal Section.
Move to next position for expansion joint
concrete
expansion joint.   bitumen felt.

Cross Section.
Levelling a cross-fall.
25mm block

3 metres   1·5 metres.

25mm×75mm

pegs at 1m. intervals.

2　　Make the formwork of old timber, 25mm x 75mm. Hammer in pegs at 1m intervals and at joins in the formwork. Allow a cross-fall of 1 in 60 across the path by setting the formwork lower on one side (25mm on a path 1.5m wide).

3　　Put in an expansion joint every 3m, using a piece of wood. The wood can either be left permanently in place, or can be replaced by a piece of bitumen felt as the concrete is laid, and the wood moved on to make the next expansion joint.

4　　Pour the concrete from the barrow into the formwork, and spread evenly using rakes and shovels to about 25mm above the top of the formwork to allow for compaction.

5　　Fill to the next expansion joint, and then tamp using a tamping beam. Move the beam forward about half its thickness each time it is dropped. Tamp from one end to the other and then repeat once.

6　　Brush with a stiff broom before the concrete has set, to give a rough surface. This will be anything from an hour to a day later, depending on the weather. After brushing, cover the concrete with wet sacking and leave for several days. Water the sacking if the weather is dry.

Concrete ramps

These can be made in the way described above, on slopes of up to 1 in 5. It is technically possible to build long sections in this way, but does not make a very comfortable path to walk on. Sections over 5m long and 1 in 7 or steeper are better built as steps. Special provision should be made

for disabled access. See Countryside Commission Advisory Series No. 15 (1981).

Roughly fill the formwork to about 25mm below the top and allow to set for a few hours. Then working from the bottom upwards, fill to the finished level. As you proceed the bottom part begins to set and prevents the surface from slumping.

Concrete sandbags

The following method was used at Bearsted, Kent, on a shady path through damp ground which could not easily be drained. Sandbags were filled with aggregate and cement in the proportion of about 4:1, and laid dry on the surface of the path. Each sandbag was tamped down with a punner, and care was taken not to tread on the bags while they were setting. No drains were dug.

The sandbags set to form a solid causeway which was above the normal water table, and not damaged by flood water. The humped surface of the concrete makes it very unobtrusive, and gives a certain amount of grip even when wet.

A wider path could be made by laying two rows of sandbags. Do not stagger the rows, as the channel allows water to drain off the path.

Riverside paths

Riverside paths liable to flooding can be strengthened by surfacing with concrete. See Chapter 11 for details of revetments.

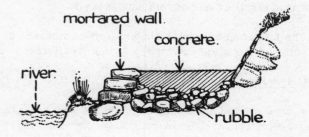

Laid Stone Paths

STONE PITCHING

This is an old method of road building, and involves the setting of selected stones into the subsoil or sand base.   The general technique is as follows:

1   Measure and mark the line of the path using pegs and string.

2   Excavate to the required depth. On stony or firm ground this will equal the long axis of the selected stone. On soft soils and peat a sub-base of sand or gravel must be laid to a depth of about 50mm, in which the stones are set.

3   Rake and thoroughly roll or tamp the sub-base to get it solid and even. Any bumps will show up in the pitching, and soft spots will sink and weaken the surface.

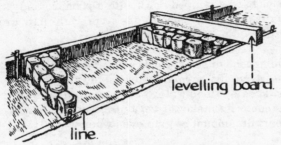

4   Set a line at the required height for the top of the pitching, and work accurately to this. This is especially important if several groups are doing pitching which must join 'invisibly'. Check across the width with a levelling board to keep the pitching even.

5   Divide the labour carefully. Like wallers, pitchers need to work singly on their own section of path. Groups of three or four can work on a section if, for example, two select and carry stone, one pitches, and one packs the gaps with broken stone.

Slopes of up to about 1 in 5 can be pitched successfully, but must be drained to prevent the path acting as a watercourse. Paths can be pitched on a cross-fall or camber as necessary.

The following examples show different methods to suit varying sites and types of stone available.

Grindsbrook Meadows, Edale

A very eroded section of the Pennine Way crossing grassland was pitched, using stone from a partly demolished wall about 50m away. The stone was transported in an Argocat (see p45), which was able to carry 450kg of stone per trip with no damage to the turf, and greatly eased the work. The soil is firm and well drained.

The path was excavated to a depth of 300mm, and dry cement was mixed with soil in the ratio of about 1:3, and spread in a layer 50mm deep to give a firm sub-base.

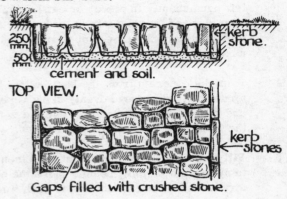

TOP VIEW.

Gaps filled with crushed stone.

It was difficult to interlock the stones satisfactorily so any gaps were filled with stones crushed with a sledgehammer. It is worth taking time to get as good a fit as possible with the pitching, as this gives a stronger finish and saves the rather tedious job of gap-filling.

Unfortunately, within a year of the pitching being completed, a new path had developed on the lower side, presumably made by walkers who found the surface uninviting and preferred to walk on the short turf. This is in rather an exceptional situation, being at the southern end of the Pennine Way, and receiving a lot of use from visitors who are not shod for walking. However, it illustrates the principle that surfacing will not be successful unless it is more comfortable to walk on than the surrounds. In the other sites described below, the surrounds to the pitched paths are rough grass, rocky or wet ground. The pitching provides a firm, smooth surface by comparison, and for this reason the paths are followed by walkers.

Pyg Track, Snowdon

This is part of a long-term scheme to repair the lower part of the Pyg Track. Heavy use over many years has eroded away the grass cover, allowing water to wash away soil and loosen rocks and boulders. The 'path' is now very wide in places, and the ground heavily damaged. The aim of the pitching is to protect the mountain from further damage, and not to make walking easier.

The line of the path is excavated using picks and crowbars to loosen rocks and heavy clay. No sub-base is needed for the pitching as the ground is very stony.   The pitching is done in sections behind a step, formed either by an existing boulder or one moved into place.   Gaps between the pitched stones are filled wtih broken stone and earth, beaten in with a walling hammer.

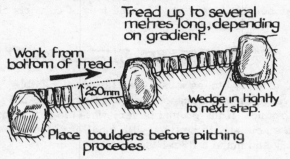

Side drains and culverts are built where necessary to keep water off the path.

Where the path cuts across a slope, the lower side is contained with a large edging stone or retaining wall.

CROSS SECTIONS.

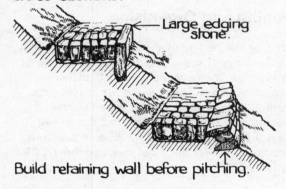

Build retaining wall before pitching.

Beacon Fell Country Park, Lancashire

Beacon Fell is acidic grassland and heath with conifer plantations, and rises to 266m above sea level.   It is a very popular viewpoint, and the turf on paths has been mostly destroyed, exposing the peat and underlying gritstone.   A variety of types of pitching have been done.

Setts are stones especially worked as a surfacing material.   They may give an 'urban' look to the path, but are easier to lay and make a smoother surface than unworked stone.   Second-hand setts from old pavements and roads may be available. Lay the setts on a 75mm sub-base of sand well

tamped down, with more sand tamped in as filling between the setts.   It is not worth trying to construct an 'informal' edge, as it is then difficult to hold the setts in place.   In the method below, used at Beacon Fell, the stone edging forms part of the side drain.

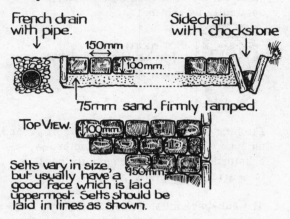

The local walling stone is a rather angular and block-shaped gritstone, heavy enough to be laid with the long axis forming the walking surface. This is only possible with stones of the dimensions shown below; anything smaller must be laid with the long axis vertical.

Another section of path was pitched using the very laborious method of selecting the largest stones from a 20 ton load of mixed grade.   In spite of the largest (about 75mm x 100mm) being too small for easy laying, the result was very good, and the path is in excellent condition after five years of heavy use.   The waste stone was laid and rolled to form the sub-base.   The time taken to complete 40m x 1½m of path was as follows. Bear in mind that this is a slower rate than would be expected using larger pitching stone.

Job	Man/days
Sorting and transporting stone	10
Removing peat	4
Laying and rolling waste stone	2
Pitching	84
Brushing sand into gaps	1
Total	101

In conclusion, pitching can be an excellent way of building paths that, while in character with the countryside, are strong enough to withstand heavy use and steep gradients.   It may also be a useful solution to erosion on short sloping sections of otherwise unsurfaced paths.   Pitching is a time-consuming task, but should give satisfying results if the following points are noted.

a   Suitable sized stone must be available. The shape and size shown below is about the optimum for unworked stone. Lay it vertically as shown.

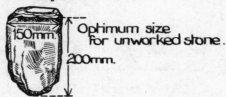

Optimum size for unworked stone.

b   Pitching is not possible on very soft ground, as even with a sub-base or membrane, subsidence will occur eventually causing the surface to collapse.

c   The sub-base must be rammed firm, and the stones wedged tight against each other so they cannot move underfoot. Any gaps must be filled with broken stone or sand. Edging stones will normally be necessary.

The effect can be of instant antiquity; even whilst making the sandstone path described above, volunteers were asked if it was a Roman artifact!

MACADAM

Like pitching, this is an old method of road-building. Stones are dumped along the line of the path, and pounded until the fragments bind to form a resistant surface.

This method was used recently at Grindsbrook Meadows, Edale, immediately beyond the trial pitched section described above. Walling stone was broken up using 10lb and 14lb sledgehammers. With this particular stone (gritstone) there was no problem with chips flying, but this might be a problem with other types of stone. A 150mm sub-base of macadam was made, then topped with a base of 20mm grade stone, and a surface of fines, tamped with a vibrating plate and punners. The rock breaking was very hard work, and the whole job took as long as a similar length of stone pitching, and was considered a much less enjoyable task. The surface is standing up well to wear.

A similar technique has been used with slate waste on a path alongside the River Teifi near Cardigan. The only problem in handling and breaking the slates was in shovelling them into a barrow for transport, as the slates could easily slip along the handle of the shovel and cut the hand. It made an effective path, but noisy to walk on.

STONE CAUSEWAY

Flat stones simply laid on the ground like 'stepping stones' are not sufficient, as they will move underfoot and sink in the wet ground. Flat stones are better used for the tops of culverts of for lining drains.

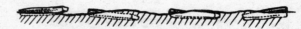

Causeway stones must be as large as possible, wedge or block shaped, and set firmly into the ground. If the ground is very wet, the bottom of the hole should be lined with smaller stones to try and make a firm base.

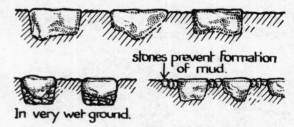

stones prevent formation of mud.

In very wet ground.

Small stones can also be packed and tamped in between the blocks. This helps strengthen the surface, and prevents the formation and spread of mud. This is particularly useful at points which get a lot of wear, such as around stiles, gateways and information boards.

Planning and Organising

Surfacing can be one of the least enjoyable jobs for volunteers if the wrong quantity or grade of material has been ordered, or if inadequate arrangements have been made for moving the material. Most tasks can be saved by the application of some ingenuity, but the wrong material in the wrong place with no transport is enough to defeat the most resourceful BTCV task leader!

Materials and transport

a   Estimation of quantities is difficult, but with care a reasonably accurate estimate can be made. Decide on the width of the path, the required depth of material, and multiply by the length of path. For example, a 1.5m width path, surfaced to a depth of 100mm, and 80m in length requires:

1.5m x .1m x 80m = 12 cubic metres of surfacing material

Some material is provided by the cubic metre, and some by the tonne. The supplier will be able to tell you how many tonnes of a certain material make up a cubic metre.

b   If possible, go to the quarry or supplier to check the grade and quality of the material. Once the lorry has dumped it, it is too late to complain!

c   Transport on site is often a major problem. Try a few estimations first.

2 shovels-full = 1 bucket
bucket (2 gal) = 0.009 cubic metre
wheelbarrow = 0.07 cubic metre of aggregate piled up, and only moveable by a strong person
         = 0.04 cubic metres of wet concrete
dumper truck = various sizes ($\frac{1}{2}$, 1, 2, 3 cubic metres)

Taking the example above, 12 cubic metres of surfacing material could be moved by 2666 shovels, 1333 buckets, 171 wheelbarrows or 24 small dumpers!

d   Consider the use of machines, such as dumper trucks, argocats and even helicopters (see p45). However, do not rule out the traditional source of supply from borrow pits (see p80), so avoiding the need for transport.

e   It may be possible for material to be dumped by machine alongside the path before the volunteers arrive. This not only reduces the chance of accident, but it may be easier to arrange for contractors to do the work at a time that suits them. Surfacing material is usually needed on soft ground, which adds to the transport problem. It may be possible to take advantage of a dry spell to shift the material by machine in preparation for a future task.

f   If material has to be moved by hand, make it as easy as you can by providing builders' wheelbarrows, in good condition, with pneumatic tyres. Keep the bearings well greased. On soft ground, provide solid wooden barrow-boards or matting such as Terram to support the barrows.

If material has to be carried, provide plenty of strong rubber or canvas buckets, or fertiliser bags. To take the strain off the arms, pack-frames or 'yokes' can be used. See also page 132.

Organising volunteers

For a leader faced with shifting a large amount of material by hand:

a   Set an easy target for the first morning as an encouragement.

b   If there are several chains or teams of people, try and make sure that the members of each team are of about the same ability so that frustrations do not occur.

c   If there has to be a long chain of volunteers, either carrying or pushing material, keep a close eye on it so that it works smoothly. Give the weaker volunteers a shorter or easier section, and put the strongest ones on the steep, soft or rough ground.

d   Where the job looks like being very tedious, break it up with, for example, five minutes rest and chat every fifteen minutes. It is frustrating and inefficient if the chain starts to break up through boredom or exhaustion.

d   Songs and the odd race can help turn a monotonous task into a memorable event!

9 Boardwalks and Bridges

The following terms are used to describe the parts of boardwalks and bridges.

Boardwalks

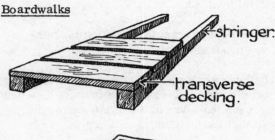

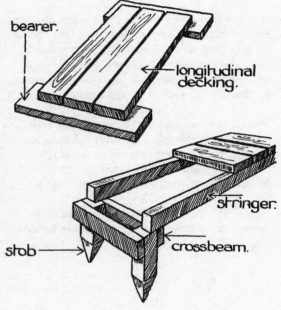

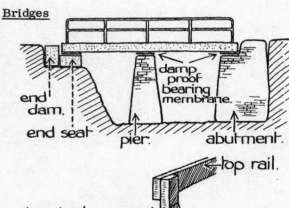

Bridges

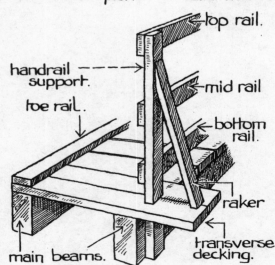

Materials and Fixings

LOCAL MATERIALS

In some situations it may be possible to use locally occurring materials, which will save on purchase and transport costs. Local timber supplies should be assessed carefully to decide whether this saving justifies a probable shorter useful working life, compared to purchased and pressure treated timber.

Uses of local materials

	BOARDWALKS	BRIDGES
Timber	Bearers Stringers Decking	Main beams - -
Stone	Boulder bearers - -	Abutments Piers Scour protection
Aggregate	Boardwalk approach	Concreting
Water	-	Concreting

Do not use sea sand or shingle for concreting, as the salt reduces the setting ability.

TIMBER

Appendix C lists different types of timber, with suggested uses and preservative treatments.

FIXINGS

Nails

Use galvanised, sherardised or zinc or cadmium plated nails.

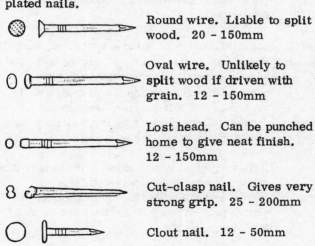

Round wire. Liable to split wood. 20 - 150mm

Oval wire. Unlikely to split wood if driven with grain. 12 - 150mm

Lost head. Can be punched home to give neat finish. 12 - 150mm

Cut-clasp nail. Gives very strong grip. 25 - 200mm

Clout nail. 12 - 50mm

Staple. Various sizes.

To reduce the chance of splitting, preferably pre-drill for nails in hardwood and larch. Always pre-drill for nails 100mm or longer. Pre-drilled holes should be 0.8 x diameter of the nail. Nails should be of the lengths shown below.

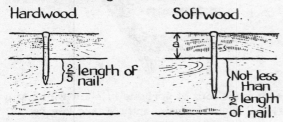

The minimum spacing of nails should be as shown below.

d = diameter of nail

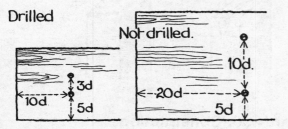

Screws

Use sheradised or galvanised screws.

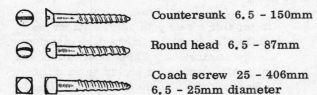

Countersunk 6.5 - 150mm

Round head 6.5 - 87mm

Coach screw 25 - 406mm
6.5 - 25mm diameter

Minimum spacing should be as shown below.

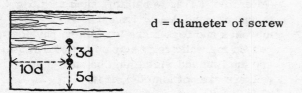

d = diameter of screw

The pilot hole should be half the screw length if fixing to softwood, and slightly deeper for hardwood. Make the pilot hole a smaller diameter in softwoods than in hardwoods. Dip screws in linseed oil or vaseline before use.

Bolts

Electro-plated bolts are the best quality.

Coach bolt. Square collar locks in wood as nut is tightened. Up to 500mm long, 5 - 19mm diameter.

Rag bolt. Ragged end holds in concrete. 30 - 200mm long, 4 - 60mm diameter.

Length of bolt must equal the thickness of timbers, two washers and nut plus 5mm.

Washers

Use galvanised or sherardised washers.

Internal diameter should be slightly larger than bolt.

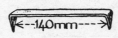

Timber connectors, used to strengthen grip between softwoods.

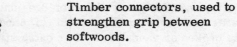

Dog clip, for joining adjacent timbers.

TIMBER JOINTS

a   Cut the joints accurately. Gaps will weaken the joint and collect water which hastens rot.

b   Always nail lighter timber to heavy. Avoid nailing twice into the same grain line.

c   Pre-drill for nails in hardwood or larch.

d   Soak all joints with preservative before assembly.

T joint

Used for bridge and boardwalk decking. Nail obliquely, preferably with the deckboard overlapping the beam (see p92).

Place deck boards with heart down.

Lap joints

The full lap joint can be used in boardwalks to fix stringer to stob. Either joint is suitable for stile rails.

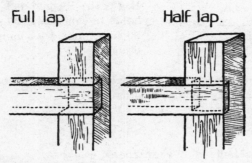

Full lap   Half lap.

Mortise and tenon

The thickness of the tenon should not exceed one third the thickness of the rail. This is a strong joint, used for handrails on bridges and bars on stiles. It is however, difficult to mend if the rail is broken or vandalised. For extra strength, the joint can be dowelled (see p150).

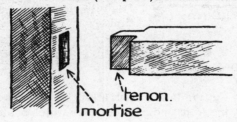

tenon.
mortise

Lengthening joints

These are used for handrails. If there is more than one rail, position the lengthening joints so there is only one on any handrail support.

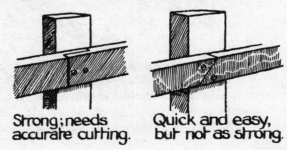

Strong; needs accurate cutting.

Quick and easy, but not as strong.

Three-way joints

These are used for boardwalk corners and steps.

Finishing

To give a neat finish, chamfer the ends of the deck boards with a surform.

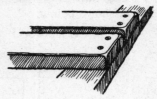

Weather the tops of uprights to allow water to run off, and chamfer edges and corners to give a smooth and attractive top.

chamfer edges.

Boardwalk Construction

Boardwalks are usually constructed for either of the two following reasons, which in turn affect their design.

a To provide a safe path across otherwise impassable terrain such as deep marsh or alongside streams. These boardwalks are obligatory, and there should be no difficulty in keeping users on the boardwalk. Handrails may be needed for safety.

b To protect a fragile habitat, such as a bog, marsh or sand dune. Except in very wet marshes and bogs there is often no physical reason for walkers to stay on the boardwalk, and the line and width must be such that walkers are not tempted off it.

Obligatory   Voluntary.

Wish-ways may develop.

Width

The required width depends on whether the boardwalk is designed to be one or two-way, obligatory or voluntary, and on the numbers of people likely to be using it at any one time. Suggested widths are between 750mm and 900mm for one-way use, and between 900mm and 1200mm for two-way use. A boardwalk receiving occasional use need only be 750mm wide, whereas a popular one-way nature trail should be at least 900mm, to allow people to stop or overtake. Heavily used boardwalks near information centres, or linking car-parks and beaches must be at least 1200mm wide. The widths for wheelchair use are given on page

Boardwalks that are designed mainly as nature trails will require large 'passing places', maybe with information boards, where groups can gather. The following design has been used at Hardwick Hall, Durham, for a nature trail through carr. On this and other nature trails, one-way use is encouraged which makes traffic flows much easier to manage.

PLAN.

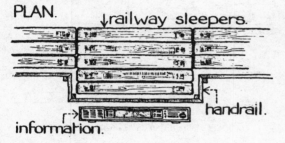

Short (ie under about 10m), single-width, obligatory boardwalks may be successful, but the path will need surfacing at either end where people gather to wait their turn.

Obligatory.

<----- Surface.

Several very effective single-width boardwalks have been built on the Dorset Coast Path between Swanage and St Aldhelm's Head, to cross springs that seep through a clay topsoil. In summer when the path is heavily used the springs are almost dry and the boardwalks are not needed. In the

winter when the ground is waterlogged and unpleasant to cross, the boardwalks become 'obligatory', and are sufficient to cope with the smaller numbers of walkers.

Single-width voluntary boardwalks are not usually worth constructing. If trampling pressure is such that a boardwalk is needed, a single-width boardwalk will not be enough to contain it.

Voluntary.

Stobs or bearers

Stob construction is usually better than using bearers as less timber is required and levelling is easier. However, bearers will be required in very soft ground in which stobs sink. Test by knocking in a few trial stobs before deciding.

Bearers can be made cheaply of local timber or boulders, if available. Timber bearers are not suitable in ground where flooding occurs, as the boardwalk will be washed away. Boardwalks with stringers laid directly on the ground are not usually successful unless very heavy, as they are difficult to level and tend to move underfoot. This also applies to pallets, which are only useful for temporary boardwalks or those on sand dunes, where blown sand holds them in place.

Foundations

If a very high standard job is required, for example linking a car-park and an information centre, it may be necessary to dig foundations to make a sufficiently solid base. At Risley Moss, Warrington, foundations were dug at 1.6m centres in a double line to support a prefabricated boardwalk. When measuring centres, always measure each time from the beginning to avoid accumulating any error.

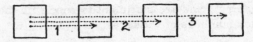

Trampling during the excavation phase may alter the levels, so all the foundations must be re-levelled before the boardwalk is put in position.

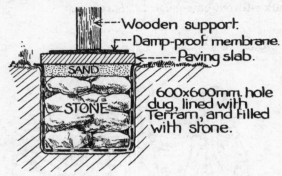

Wooden support.
Damp-proof membrane.
Paving slab.
SAND
STONE
600x600mm. hole dug, lined with Terram, and filled with stone.

Decking

Most boardwalks have decking of sawn timber, laid transversely, with a gap of about 25mm. Boardwalks designed for use by the disabled or people in wheelchairs should have a maximum gap of 10mm on transverse decking, and no gap on longitudinal decking.

Transverse deckboards can either be attached flush with the sides of the stringers, or with an overlap.

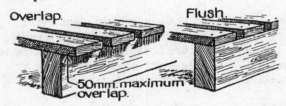

Overlap. Flush.
50mm. maximum overlap.

A flush finish gives a neater appearance and the boards are less easy for vandals to remove, but the overlap is a stronger and structurally more efficient construction, as for a given finished width, the unsupported span is smaller.

The overlap method has the further advantage that the deckboards are less likely to split during construction, and the boards will have a longer useful life before rotting from the ends causes collapse. The overlap should not be greater than 50mm. Whilst attaching the deckboards, use a template the size of the overlap to ensure a symmetrical finish.

Pre-drilling

Pre-drilling of deckboards on site is not usually practical because it takes too long. However, as pre-drilling does reduce the possibility of the deckboards splitting, it may be worth doing this

in a workshop before taking the materials on site.

Annular nails also reduce the chance of splitting, and being difficult to remove, are a deterrent to vandals.

Non-slip surfaces

One of the problems with boardwalks in tall or enclosed vegetation such as carr is that they become very slippery, as algae grow in the damp conditions. As these are the types of boardwalks that tend to be constructed for nature trails and teaching purposes, safety considerations are particularly important. There should always be a gap between deck boards to allow water to drain through, and a slightly wider gap than is usually recommended may give more grip, as the shoe bends into the gap in the decking.

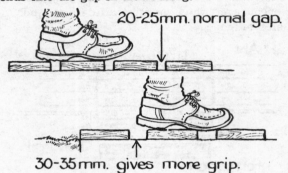

20-25mm. normal gap.

30-35mm. gives more grip.

For the same reason, boardwalks of split timber give better grip in wet conditions than does smooth sawn timber.

Battens of rough timber can be tacked across half or the full width of the decking. Battens cut from fruit and vegetable boxes (discarded by greengrocers) are ideal, being of thin but rough wood which gives grip.

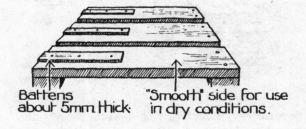

Battens about 5mm. thick. "Smooth" side for use in dry conditions.

Other methods tried are to staple 'Netlon' plastic-covered wire netting in a strip down the centre of the boardwalk, or to cover the entire decking with chicken wire.

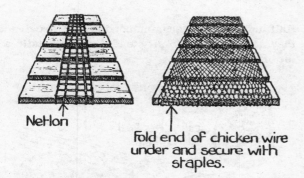

Netlon

Fold end of chicken wire under and secure with staples.

Railway sleepers can be a problem as they may be slippery in wet or frosty weather, and tarry in hot weather. Epoxy tar can be sprayed or painted onto the sleepers, and then spread with grit. This will need renewing periodically. Alternatively, grit can simply be sprinkled direct onto the sleepers every few months or as necessary.

Handrails and toeboards

Handrails may be fitted for safety or to keep walkers on the boardwalk. They must always be strong enough to bear the weight of people leaning against them. Timber sizes are given on page 107.

Toeboards give some security for wheelchair users and for people who have a fear of walking on raised surfaces. They also discourage vandalism to the decking.

Prefabrication

Some prefabrication may be helpful, particularly if the weather is bad. It also means less duplication of tools. However, it does have its problems in that sections can become very heavy and awkward to carry, and difficult to level in 'non standard' situations. Its best use is for boardwalks without stobs, when the stringers and cross-beams can be assembled on a jig, and then the decking attached once the frame is in position (see p97).

Line and profile

Boardwalks in woodland, carr or reedswamp are usually hidden by vegetation, but those across open ground need careful siting and design.

Avoid a straight line.

For reasons of cost, the line taken is likely to be the shortest route possible. The appearance can be improved by one or two gentle corners. Curves are difficult to construct except with sleepers laid directly onto the ground.

Construction of attractive boardwalks on sloping ground is less easy, but try to avoid a very raised profile, particularly if very heavy timbers such as railway sleepers are used.

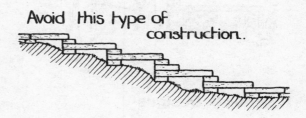

Avoid this type of construction.

Steps

Board in the risers of steps to neaten the appearance and protect the stringer ends.

Flights of steps can be made easier to distinguish and so safer by alternating the decking between transverse and longitudinal. A slope can then be turned with the longitudinal decking, giving a very attractive result.

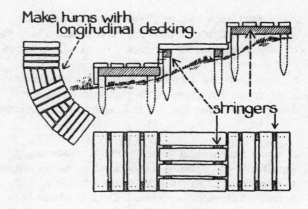

Make turns with longitudinal decking.

stringers

93

The Countryside Commission for Scotland (CCS Information Sheet 6.11) recommend that all steps be constructed of longitudinal decking because this allows off-cuts to be used, and is more durable because the end-grain takes most of the wear.

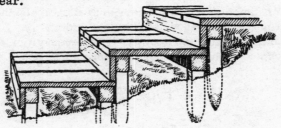

Half logs are not recommended for the decking as they need notching to sit steady, which entails a lot of work.

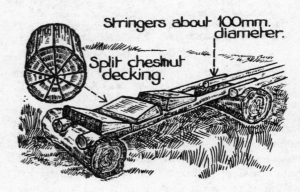

Stringers about 100mm. diameter.

Split chestnut decking.

Boardwalk Designs

PUNCHEON

This is an American term to describe a structure made of local timber.

Each section of puncheon requires two trees; one stringer and one bearer being cut from each tree.

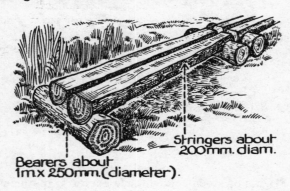

Stringers about 200mm. diam.

Bearers about 1m x 250mm. (diameter).

1 Make a flat surface either by scoring the stringer with an axe and chipping away the scored wood, or cut away the top using a chain saw.

2 Seat each bearer solidly in a trench, and cut notches to take the stringers. Place the stringers with a gap of no more than 50mm.

3 Attach the stringers to the bearers using large cut-clasp nails, or 6mm diameter reinforcing rod cut to length and hammered in.

A more economical version can be made using smaller logs. The boardwalk shown below, made in chestnut, has lasted 10 years without major repair at Slapton Ley, South Devon, where it is used to cross an area of reedswamp.

RAILWAY SLEEPERS

These have been extensively used in the past few years, but are now becoming very expensive in some parts of the country as supplies dwindle. During 1981, prices from £4.00 to £14.00 per sleeper were found, so check local supplies before deciding on this method of construction.

This construction has been used at Risley Moss, Warrington. Whole sleepers were laid direct on the moss as stringers, with half sleepers used for the decking. Spacers of 10mm thick pieces of wood were used to make drainage gaps. Spacers were nailed to the side of the sleeper, and then each sleeper knocked tight against the previous one using a sledgehammer. The decking sleepers were not nailed to the stringers, as the weight is sufficient to hold them in place.

Finally, galvanised fencing wire was pulled taut along the decking and stapled to each sleeper, to discourage vandals dislodging the decking.

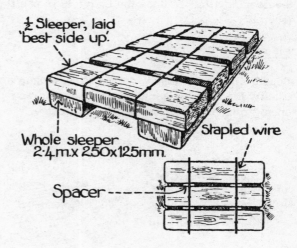

½ Sleeper, laid 'best side up'.

Whole sleeper 2.4m x 250x125mm.

Stapled wire

Spacer

Corners were turned in either of the two ways shown below.

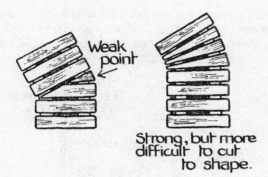

Weak point

Strong, but more difficult to cut to shape.

Sleepers can be more economically used as longitudinal decking, but boardwalks of only two sleepers width are not recommended except for short obligatory sections. Longer sections inevitably develop paths alongside and look ugly in proportion, being much too narrow for the depth of timber. Three sleeper width boardwalks are sufficient for two-way traffic, and appear to 'sit' more attractively in the landscape.

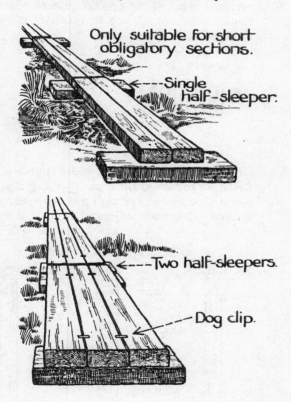

Only suitable for short obligatory sections.

Single half-sleeper.

Two half-sleepers.

Dog clip.

Boardwalks made entirely of sleepers are relatively quick and simple to construct, provided that the sleepers do not have to be carried too far.

SLEEPER AND SAWN TIMBER

a   This takes longer to construct than the preceding design, but saves on sleepers and is less obtrusive in short vegetation. The stobs and cross-beams are 100mm x 100mm which minimises wastage. A boardwalk of this type has received moderate two-way use at Ben Lawers, Perthshire, for over seven years. The thick growth of vegetation at the side and under the edges of the boardwalk are testimony to its success.

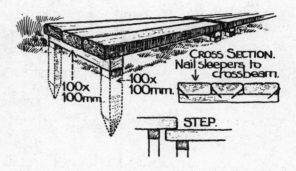

CROSS SECTION. Nail sleepers to crossbeam.

100 x 100mm.

100 x 100mm.

STEP.

b   The design below is suitable for softer ground, and has been in use for many years at Malham Tarn, Yorkshire. Originally creosoted timber was used for the decking, but this did not prove durable as the creosote was not applied under pressure. Tanalised elm is now used. In very wet sections the sleeper bearers are turned to give extra height.

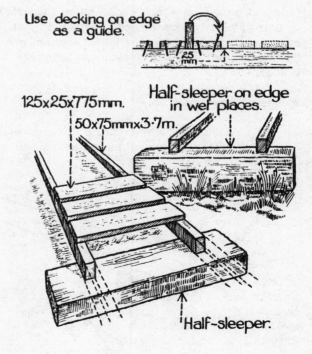

Use decking on edge as a guide.

25 mm

125x25x775mm.

50x75mm x 3.7m.

Half-sleeper on edge in wet places.

Half-sleeper.

c   The following design has been in use for about twenty years at Hothfield Bog, Kent, where the wet conditions and possible flooding require the boardwalk to be raised up to 500mm above the ground. The design incorporates a handrail on one side.

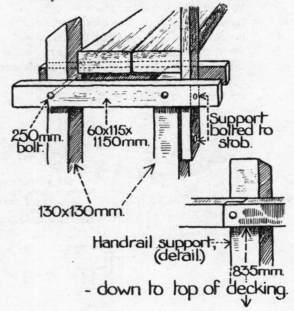

250mm. bolt.

60x115x 1150mm.

Support bolted to stob.

130x130mm.

Handrail support. (detail.)

835mm.

- down to top of decking.

SAWN TIMBER

The spacing of the decking and the dimensions of the timber give these boardwalks a 'lighter' appearance in the landscape, than those involving the use of railway sleepers. They can be very durable if good quality, properly preserved timber is used, and the construction is sound and level. Most sawn timber boardwalks are built in independent sections, with each section supported by four stobs. In firm ground, where each stob can take a greater pressure, joined sections can be constructed, which is quicker and uses less timber.

Independent sections

a   This design has a half-joint which is strong and resists any lateral movement caused by uneven ground.

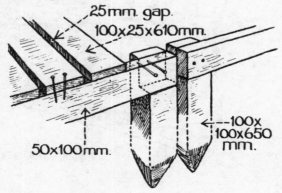

25mm. gap.

100x25x610mm.

50x100mm.

100x 100x650 mm.

It is also fairly vandal-proof, as the stringers cannot be removed without first taking off all the decking.

The dimensions given are of a boardwalk at Ben Lawers, Perthshire, which forms part of a one-way nature trail. Corners are turned by one of the following methods.

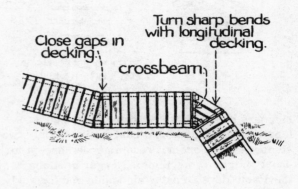

Close gaps in decking.

Turn sharp bends with longitudinal decking.

crossbeam.

In order to construct a level section across a slope, first knock in stobs A and B, leaving just over 100mm protruding to lift the stringer clear of the ground. Level A and B. Repeat with stobs C and D, levelling to the stringer already in place.

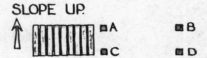

SLOPE UP.

A   B

C   D

b   The design below is for a wider boardwalk, using larger sized timber. By using the same size timber for stob and stringer, a simple T joint can be made.

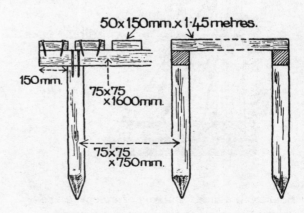

50x150mm.x 1.45 metres.

150mm.

75x75 x1600mm.

75x75 x750mm.

The T joint is not as strong as the half-joint shown above, as it gives no resistance to lateral movement, and the fixing is into the end grain which does not grip as well.

A step can be made as shown.

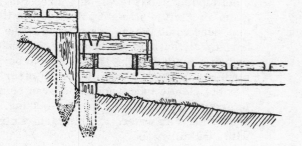

Either of the methods shown below can be used for making steps.

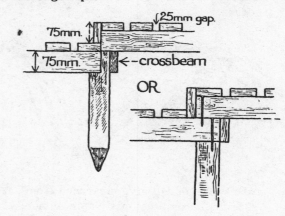

c   Instead of stobs, this boardwalk is supported by boulders. Cross-beams are necessary to give the structure rigidity. The frame of stringers and cross-beams can be quickly constructed on a jig.

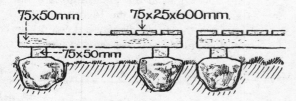

Jig made of blocks, nailed to thick plywood. This also gives solid base, for work in marshy or muddy conditions.

Joined sections

Although this boardwalk has half the number of stobs of the designs shown above, it requires more timber for the cross-beams and middle stringer. It is useful for rocky and stony ground, where placing each stob is hard work, but gives a firm support. This example is at Craigellachie, Aviemore, where the boardwalk climbs a rocky hillside with fragile plant communities.

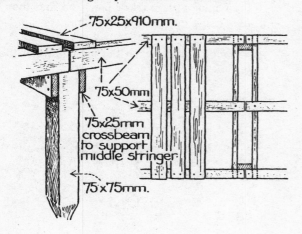

Footbridges

This section discusses the design and construction of footbridges under nine metres in length. Any bridge longer than this should be designed by an engineer. Also consult on the construction of bridges under nine metres if there is a high risk to safety or doubtful ground conditions.

A major source of information on this subject, used for parts of this section, is 'Footbridges in the Countryside' (Countryside Commission for Scotland, 1981).

Consents

a   Consent from the Highway Authority for a bridge on a public right of way.

b   Consent from the Regional Water Authority (England and Wales) or River Purification Board (Scotland), plus anyone with navigation rights.

c   The Town and Country Planning (Scotland) Acts apply to footbridges, which may need planning permission.

d   The Health and Safety at Work Executive have the right to inspect to ensure that maintenance work is being done correctly.

e   The owner's insurers may need to be advised.

Site selection

In the choice of site, look for the following:-

a   Shortest span.

b   Solid banks which will provide strong

foundations. Test with a soil auger or by digging test pits. Avoid bends where erosion occurs.

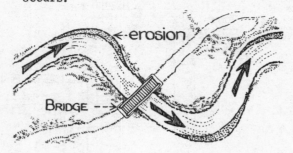

c   Banks which will give clearance from flooding. The Regional Water Authority (England and Wales) or River Purification Board (Scotland) will advise on flood levels.

d   A site which fits the desire line of the path.

e   Easy access for plant and materials.

Survey

1   Section across gap. Preferably this should be surveyed using a Dumpy level or similar. If this is not possible, use the following method.

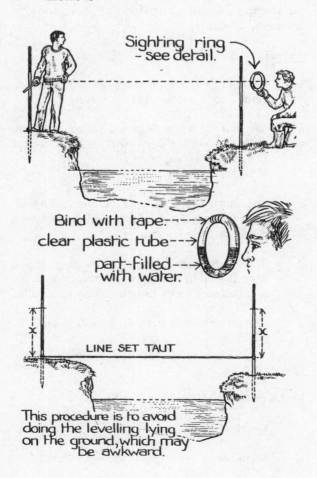

Sighting ring - see detail.

Bind with tape.
clear plastic tube
part-filled with water.

LINE SET TAUT

This procedure is to avoid doing the levelling lying on the ground, which may be awkward.

You will need a surveyor's tape or line, a steel tape, a spirit level, and three ranging poles or similar straight poles. You will also need some means of setting the line horizontal, such as a clinometer, sextant or home-made level. The latter can be made out of about 600mm length of clear flexible plastic tube such as caravan plumb-in or siphon tubing used in wine-making. Partly fill with water, and bind into a circle with insulating tape.

Set a ranging pole firmly in the ground on either side of the gap. Check that they are vertical using the spirit level. One person then uses the levelling instrument against a fixed mark on the nearer pole while another person moves a marker against the further pole according to the instructions of the first person. Mark the level, and then measure the same distance down (x) on both ranging poles, and set the tape or line taut.

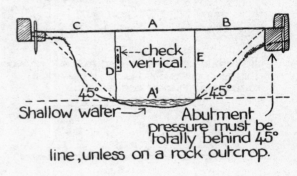

Shallow water → Abutment pressure must be totally behind 45° line, unless on a rock outcrop.

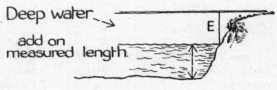

Deep water → add on measured length.

Using a ranging pole set vertically, measure heights D and E, and read off or mark distances C, A and B. Measure A^1 as a check. Replace the two ranging poles on the banks with marker posts so that the exact line can be found again.

2   Sketch or photograph the elevations of the banks.

ELEVATION.
marker post
stream.

98

3   Draw a sketch map showing details of the stream bed such as shoals or points where erosion is occurring. Include bank features and obstacles such as trees or rocks.

4   Note the nearest vehicle access.

5   Look for evidence of flood such as twigs and debris in bankside trees, and deposits of river gravel. Note the height of these above the marker post.

6   Consider whether or not the bridge should be a 'feature' in the landscape. In flat agricultural areas with deep drainage ditches, handrails of bridges provide useful waymarks.

In scenic or historic landscapes it may be desirable to site the bridge out of view.

In less interesting scenes, an unusual bridge can provide a focus.

Loading

The loading on a bridge includes the following:

a   The dead load of the bridge structure.

b   The weight of bridge users plus the dynamic effect of their movement. The forces exerted by users leaning or falling against the handrail. Use is calculated for either normal or crowd loading, as shown.

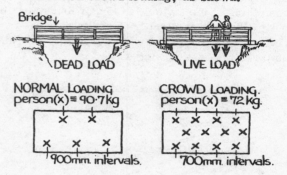

c   The horizontal effect of the wind on the bridge and the user, and the vertical effect from suction or pressure on the decking.

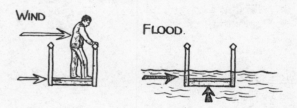

d   The vertical pressure from weight of lying snow and ice.

e   Water pressure on the bridge during flood, plus pressure from floating debris.

These factors are given for interest only, as specialised knowledge is needed to design a structure for the calculated loadings. All the designs given in this chapter are suitable for normal pedestrian loading. Any bridge that may attract crowd loading, for example near a car-park used by coaches, and bridges on bridleways, should be designed by an engineer.

Types of Bridges

The following illustrations show various types of bridges suitable for rural paths. Construction details are given later in the chapter. The dimensions of timbers are not given as these must be calculated for each site from the information given on page 106.

SIMPLE BRIDGES

These are low bridges without handrails, suitable for situations where there is no risk to safety.

Clapper

Stone clapper bridges are an ancient type of bridge and can only be constructed where large enough slabs are obtainable nearby or can be brought to the site. A 20th Century equivalent which retains something of the style of a clapper bridge can be built with railway sleepers fixed to mortared piers. This is similar to a boardwalk, but raised on piers to cross flowing water.

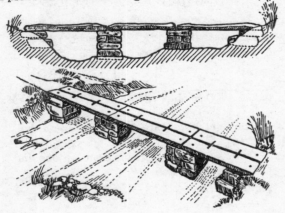

Sleeper

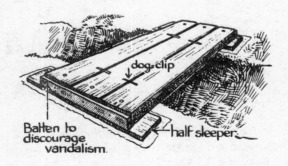

Batten to discourage vandalism.

half sleeper.

dog-clip

Duckboard

For short spans, the duckboard can be pre-fabricated. See page 105 for methods of fixing the duckboard to the half-sleepers.

half sleeper.

Toe-board; stops walkers slipping off, and discourages vandals.

BRIDGES WITH HANDRAILS

Single beam

These are suitable for footpaths in locations where use by wheelchairs is not possible. They are economical in materials, and with their simple line and 'monkey-bridge' appearance, they fit well in remote and adventurous landscapes.

Either a V or square section beam can be used. As shown below, the beam can be anchored either to a cast concrete end seat, or into an abutment.

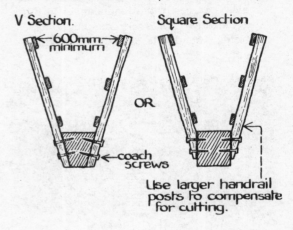

V Section.

600mm. minimum

Square Section

OR

coach screws

Use larger handrail posts to compensate for cutting.

Fill with Secalastic pre-drill.

dpm

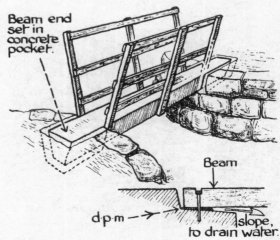

Beam end set in concrete pocket.

Beam

d.p.m

slope, to drain water.

100

Log bridge

These are best used at sites where logs are locally available (see p106). Handrails must always be raked to give a secure fixing.

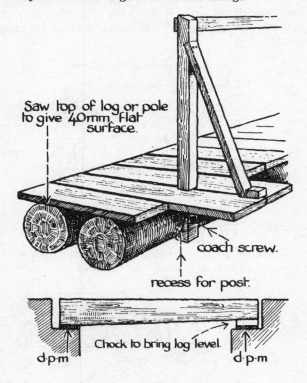

Saw top of log or pole to give 40mm flat surface.

coach screw.

recess for post.

Chock to bring log level.

d·p·m   d·p·m

Sawn timber bridge

This is the standard type of bridge for footpaths. Two or three beams are used, according to the span and the required width. Steel ties through the width of the bridge hold the handrail posts in position. Struts are skew nailed next to each tie to give the structure rigidity.

Deckboards can either be flush with the beams, as shown, or overlapped by about 50mm, and cut to fit around the handrail posts.

Nail struts next to each tie.

Abutments and End Seats

The abutments or end seats keep the beam ends dry, and provide secure anchorage to prevent their movement in use and during flood. The diagram on page 98 shows how to calculate the correct distance of the end seat from the bank.

The end dam keeps surfacing and downslope material off the beam ends.

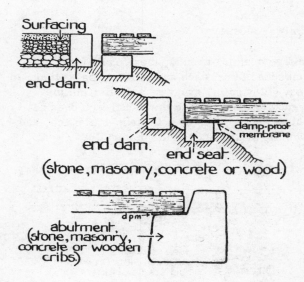

Surfacing

end-dam.

end dam.
(stone, masonry, concrete or wood.)

damp-proof membrane

end seat.

abutment,
(stone, masonry, concrete or wooden cribs.)

dpm

Various types of abutments and end seats are shown below.

Timber cribs

These can be built either of sawn timber or of logs. Apply at least three coats of creosote to any timber which will be in contact with the ground.

A common technique on American trails is to fill a log crib with large stones. The bridge beams are nailed or wired to the top timber of the crib.

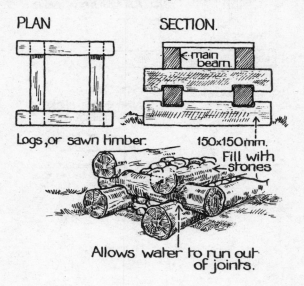

PLAN   SECTION.

main beam.

Logs, or sawn timber.   150x150mm.
Fill with stones

Allows water to run out of joints.

Three different patterns for timber cribs are
shown below.

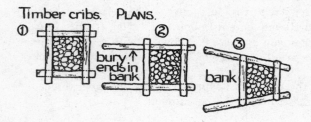

Railway sleepers

In free draining ground, these can simply be
embedded in a trench and securely backfilled. In
heavy clay soils, excavate a drain from the
trench to the ditch, so that the trench does not
act as sump. Put the half sleeper in place, and
backfill with free draining material.

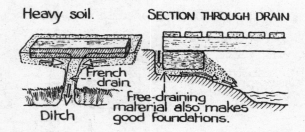

In places where flooding may occur, the sleepers
must be secured to the ground by stakes. Always
bed the sleepers on a foundation of stony material
to prevent uneven settlement.

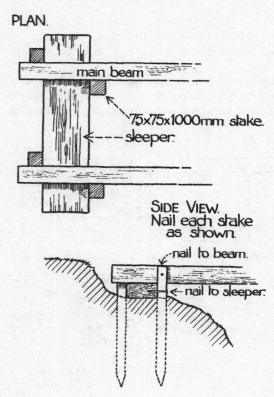

Stone and mortar

Stone abutments should be mortared to give
resistance against water erosion. Use the largest
stones for foundations and corners, and fill the
centre with rubble. Finish with a capping of
mortar and set rag bolts in position for main beam
attachment.

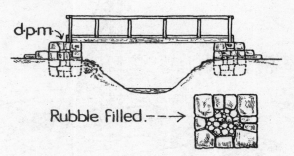

Cast concrete

A simple end seat can be made by excavating a
trench and filling it with 1:4 concrete.

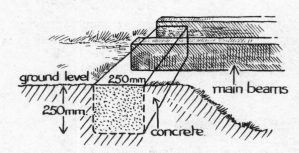

Normally a raised end seat will be required to
give sufficient clearance for the beams. This
should sit on a firm foundation of bedrock or subsoil. On deep soils where this is not possible,
construct foundations as shown below.

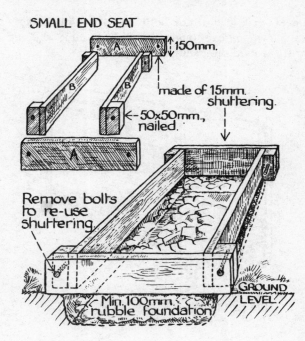

The large end seat with concrete foundations is cast in one piece. Excavate the foundation, assemble and oil the shuttering, and then suspend it as shown on two beams. Fill with 1:4 concrete. The shuttering can be removed and re-used.

LARGE END SEAT.

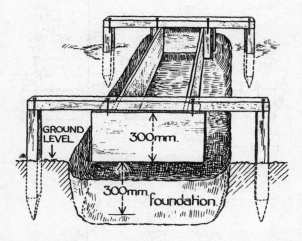

The shuttering for the end seat with ramp shown below can also be re-used. Use 15mm shuttering ply with 75mm x 50mm timber nailed along each edge to give extra strength. Drill as shown, for assembly with bolts 150mm long.

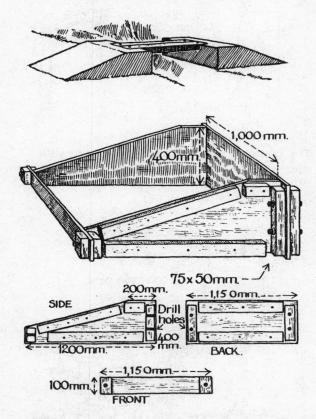

Carry shuttering to site, and set in levelled foundations about 100mm deep. Oil with shuttering oil. To save on concrete, the central part can be filled with gravel, rubble or rubbish such as cans and bottles. Leave at least 100mm margin at the sides and 150mm at the head wall to give an adequate thickness of concrete. Bottles form a very strong structure if laid in layers inside.

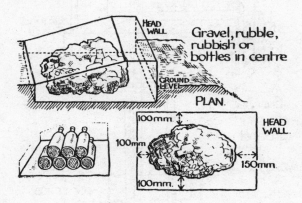

Alternatively, another box of shuttering with a top can be made. This acts as a liner and makes a hollow abutment.

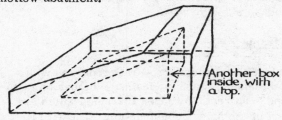

Fill with concrete in layers about 100mm thick and tamp to remove air bubbles. To give a neat finish, shuttering can be set into the head wall to form a recess for the main beams. Put rag bolts in place while concrete is still wet. Brush ramp to give rough surface before concrete has set.

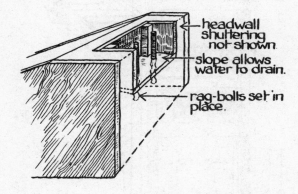

Cover with damp hessian or sacking to slow drying, and leave for 24 - 48 hours in summer, and a minimum of 48 hours in winter. Remove bolts and knock shuttering away with a mallet.

Do not concrete if frost is likely within 12 hours.

Alternatively, concrete blocks can be used to build the end seat. These are easier than lintels to carry on site, but are more obtrusive because of the mortar joints.

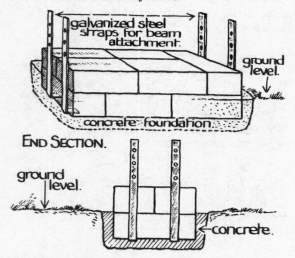

END SECTION.

Cast concrete steps require complex shuttering and concreting which is best left to an expert. If steps have to be made, build them out of concrete blocks, stone or wood.

H frame

Bridges crossing narrow watercourses with stable banks, such as drainage ditches on farmland, can rest on wooden H frames which are part of the bridge construction. These are simple and quick to construct.

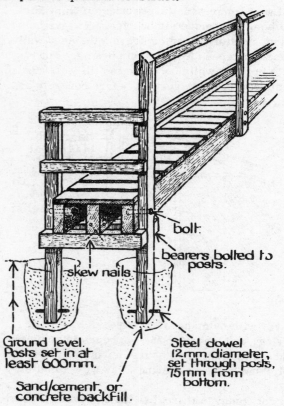

Ground level. Posts set in at least 600mm.

Sand/cement, or concrete backfill.

Steel dowel 12mm. diameter, set through posts, 75mm from bottom.

Pre-cast concrete

Concrete lintels make good end seats, being of suitable dimensions and with a fairly smooth and unobtrusive finish. Holes for the main beam fixings should be drilled by an expert, using a suitable masonry drill and safety guards. If extra clearance is needed, mortar concrete blocks onto the lintel.

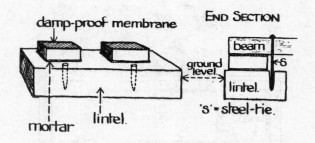

ABUTMENTS TO STREAM LEVEL

These can be made of stone, concrete blocks, cast concrete or concrete pipes.

If the stream level is high, construct a sandbag dam to keep the worksite dry. Excavate foundations to at least 450mm.

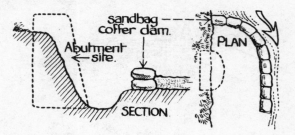

The abutment must either be rounded or with wing walls to allow the smooth flow of water. Fill abutment with rubble and cap with concrete. Set rag bolts in position.

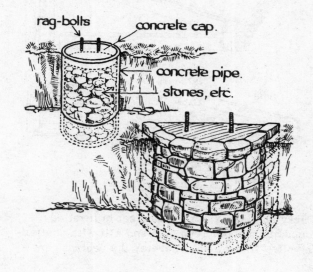

Cast concrete abutment with wing walls.

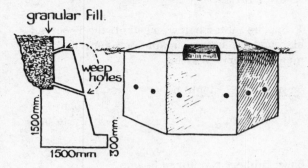

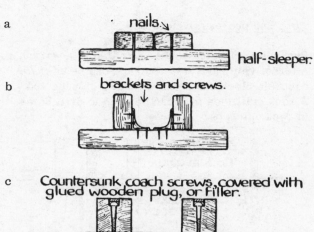

a nails.

half-sleeper.

b brackets and screws.

c Countersunk coach screws, covered with glued wooden plug, or filler.

Coach screws between decking.

PLAN.

decking

coach screw

PIERS

A pier may be necessary either to span a very wide shallow stream, or an asymmetrical stream valley. The two parts of the bridge superstructure should be independent, so that no stress is put on it if differential settling occurs. Piers can be built out of mortared stone, concrete blocks, cast concrete or concrete pipe. Designs are not given here because engineering advice will normally be necessary.

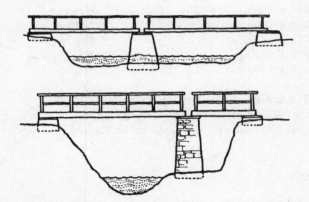

Simple log cribs can be made, but will have a limited life.

FIXING THE MAIN BEAMS

Methods of fixing the main beams to the abutment or end seat are shown below. A damp proof membrane of bitumenous felt must always be placed immediately below the main beam.

<u>Fixing to wood</u>

Use any of the three following methods.

<u>Fixing to cast concrete</u>

Screw the rag bolts through a template to ensure that the spacing is accurate.

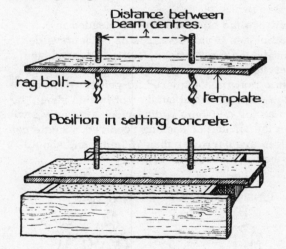

Distance between beam centres.

rag bolt.

template.

Position in setting concrete.

Logs can be secured as shown.

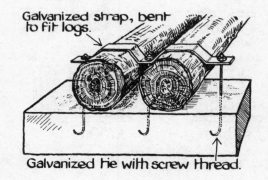

Galvanized strap, bent to fit logs.

Galvanized tie with screw thread.

105

Stone and pre-cast concrete

Pre-cast concrete should be drilled in a workshop.
Alternatively, galvanised straps can be used for
concrete block end seats as shown on page 104.
A rock drill (see p45) can be used to drill holes
in rock outcrops.

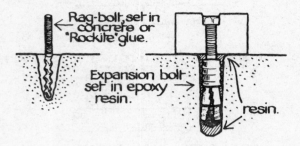

Design of Superstructure

WIDTH

The width is measured between the handrails.
When calculating the width of decking and the
distance apart of the main beams, allowance must
therefore be made for the thickness of the handrails.

Below are the standards recommended in
'Footbridges in the Countryside' (Countryside
Commission for Scotland, 1981). For bridges
receiving occasional pedestrian use a width of
750mm may be considered adequate. However,
as the same size and number of beams should be
used as for a 900mm width bridge, the only saving
is on the amount of decking required. Single beam
bridges are narrower than 750mm, and should
only be used at remote sites.

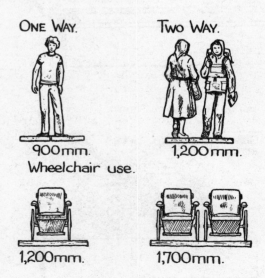

MAIN BEAMS

From the tables below, read off the required
size and numbers of main beams for the
measured bridge span.

Sawn timber: single beam

	Span
300 x 300mm	Up to 9m

Sawn timber: two/three beams

	750 - 900mm wide deck	
	No. of beams	Span
150 x 75mm	3	3 m
200 x 100mm	3	4.5 m
250 x 150mm	3	6.5 m
250 x 200mm	2	6.25 m
300 x 225mm	2	7.75 m
350 x 250mm	2	9.3 m

	1200mm wide deck	
	No. of beams	Span
150 x 75mm	3	2.75 m
200 x 100mm	3	4 m
250 x 100mm	3	5.75 m
250 x 200mm	3	7.5 m
300 x 225mm	3	8 m
350 x 250mm	3	9.6 m

Timber should be softwood Grade GS or MGS
Species S1 or S2 (see p 184), pressure
impregnated with creosote to $100Kg/m\frac{2}{3}$.

Log

Average diameter	750 - 900mm wide deck	
	No. of logs	Span
250mm	2	5 m
300mm	2	6.7 m
350mm	2	8 m
400mm	2	9.75 m

	1200mm wide deck	
	No. of logs	Span
250mm	3	5.5 m
300mm	3	7 m
350mm	3	8.5 m
400mm	3	10 m

Logs should be of Douglas fir, larch or Scots
pine. They must be straight, and of the diameter
specified for the middle third of the log length.
Remove bark and branches, but do not trim flush.
Soak all cut areas with at least three coats of
creosote, as soon as possible after felling.
Although this will not soak in far, it eliminates
fungal growth on the cut ends, and reduces end

cracking by slowing the rate of drying out.

The log should not be used if it has knots larger than 100mm, signs of fungal decay, or a marked spiral grain slope. Also reject logs that have surface cracks of 0.2mm width at time of felling (check with feeler gauge) or 3mm after seasoning. The total depth of cracks in any section must not exceed one third the diameter of the log.

Logs are best used at inaccessible sites which have suitable trees nearby. It is not worth transporting logs to a treatment plant, because of the cost, so they should only be used where a working life of about 10 years is acceptable. Telegraph poles are already treated, and will have a much longer life and are therefore worth transporting some distance.

Steel

Universal beam section	Span
152 x 89 x 17.09 Kg/M	Up to 6.25 m
203 x 133 x 25.00 Kg/M	6.25 to 8.50 m
254 x 146 x 31.00 Kg/M	8.50 to 10.40 m

For designs of steel beam footbridges see 'Footbridges in the Countryside' (Countryside Commission for Scotland, 1981).

DECKING

The gap between the deckboards should be as shown below:

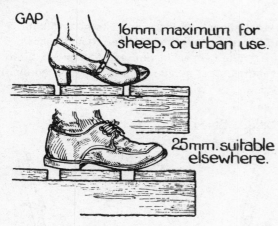

GAP

16mm maximum for sheep, or urban use.

25mm suitable elsewhere.

For normal pedestrian use of decking up to 1200mm width, either of the two following sizes of timber are suitable:

150 x 50mm softwood S1 or S2
100 x 36mm hardwood MD, D or VD (see p183).

These deckboards are thick enough to withstand

many years use before replacement becomes necessary. As explained on page 92, it is recommended that deckboards overlap the outer edge of the main beam by about 50mm. Always pre-drill, and attach using no.12 x 90mm sherardised screws. A strip of bituminous felt 20mm wider than the main beam, laid between decking and beam, will help protect the beam from rot.

HANDRAILS

Handrails must always be strong enough to withstand people leaning on or falling against them. The number of handrails will depend on the situation and the expected use.

A single handrail on one side only gives the minimum protection. In hazardous situations two or three rails on both sides should be used, with wire mesh if there is a risk to the safety of children or animals. The height should be as shown below.

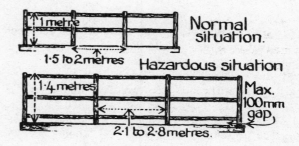

1 metre — Normal situation.
1.5 to 2 metres

Hazardous situation
1.4 metres — Max. 100mm gap
2.1 to 2.8 metres.

The spacing of the handrail posts within the limits indicated above give a pleasing proportion to the handrail, which is the most conspicuous part of the bridge. However for structural reasons, raked handrail posts should be a maximum of one metre apart.

To design the handrails:

1   Decide on cantilevered or raked fixing for the handrail posts (see over).

2   Decide on the required height and number of rails for the situation. For cantilevered posts, using the diagram above and the known total bridge span, choose a suitable spacing between handrail posts, as measured between post centres. The spacing must be consistent.

3   Using the following table, read off the required timber size in the chosen timber grade.

Spacing	Post sizes a x c		
	Softwood GS	Softwood SS	Hardwood M50
1000	100 x 75	100 x 75	100 x 75
1250	100 x 100	100 x 75	100 x 75
1500	100 x 100	100 x 100	100 x 75
1750	120 x 100	100 x 100	100 x 100
2000	120 x 100	100 x 100	100 x 100
2250	150 x 100	120 x 100	100 x 100
2500	150 x 100	120 x 100	120 x 100

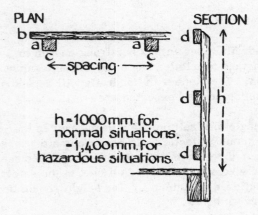

PLAN

SECTION

h = 1000 mm. for normal situations.
= 1,400 mm. for hazardous situations.

4   Using the table below, read off the required rail size for the post spacing.

Rail size b x d	Spans between posts		
	Softwood GS	Softwood SS	Hardwood M50
50 x 75	1400	1500	1750
50 x 100	1600	1700	1900
50 x 150	1800	1900	2250
75 x 50	1900	2100	2500
75 x 75	2200	2300	2700

Different methods of fixing the handrails to the posts are shown below.

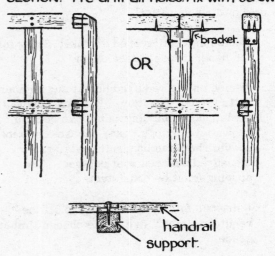

SECTION. Pre-drill all holes. Fix with screws.

bracket.

OR

handrail support.

HANDRAIL POST FIXING

Bridges with main beams of sawn timber should have cantilevered handrail posts, which are strong and simple to construct. Bridges with log or steel beams cannot easily have posts attached in a cantilever, and require a raked fixing.

Cantilever

The strongest fixing is made by using a 20mm mild steel tie, with two 76mm diameter x 6mm washers and nuts. The tie should be positioned just above the mid point of the beam. Before tightening the nuts, skew nail a 100mm x 75mm strut into position as shown.

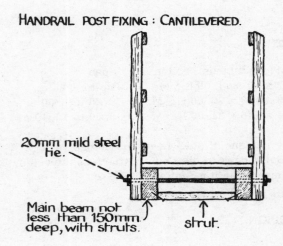

HANDRAIL POST FIXING : CANTILEVERED.

20mm mild steel tie.

Main beam not less than 150mm deep, with struts.

strut.

On single beam bridges, use 12mm x 150mm coach screws, in either of the methods shown below. The U bracket is made of 25mm x 2mm galvanised steel strap which is pre-drilled before plating.

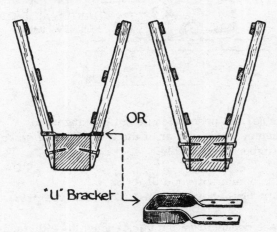

SINGLE BEAM.

OR

"U" Bracket

Raked

On log and steel main beams, the handrail posts must be supported by a raker to two deckboards extended to a distance equivalent to half the height of the handrail post.

The handrail posts should be not less than one metre apart. A wider spacing requires sizes of post and raker which will be too large to be safely supported by the extended deckboards.

The bottom end of the raker is cut to shape and tightly butted against a 100 x 50 x 300mm long stop, secured to each deckboard with two 12mm x 90mm galvanised screws in pre-drilled holes.

The handrail post and raker are joined at the top with a 12mm x 140mm galvanised bolt with two 50mm x 3mm galvanised washers.

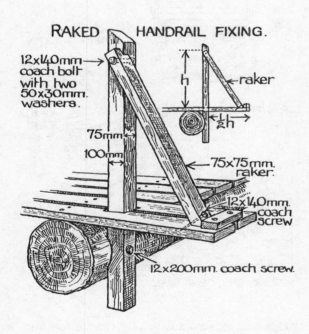

RAKED HANDRAIL FIXING.

12 x 140mm coach bolt with two 50 x 30mm washers.

75mm
100mm

h
raker
1½h

75 x 75mm. raker.

12 x 140mm. coach screw

12 x 200mm. coach screw.

TOEBOARDS

A 75mm x 50mm toeboard can be used instead of, or as well as, a handrail. It acts as a safety feature to delineate the edge, and discourages vandalism to the deckboards. Using no. 12 x 90mm sherardised screws, attach the toeboard to alternate deckboards, so that its outer edge is directly above the outer edge of the main beam.

Finishing

RESTRICTING USE

Keeping stock off and away from the abutments of bridges can be a problem. Although a bridge should be able to withstand knocks and use as a rubbing post by cattle, stock will puddle the ground around the bridge and damage the stream bank. Fencing can be used, but this is troublesome to build, requires stiles or gates, and looks rather ugly. A better solution may be to surface the ground around the bridge and revet the bank to prevent damage by puddling, and put a gate or stile on the bridge if it is necessary to keep stock off it.

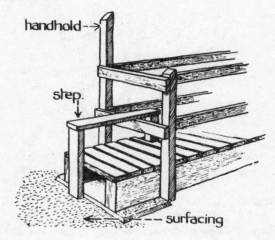

handhold
step
surfacing

The inner step support is attached to the main beam as shown below.

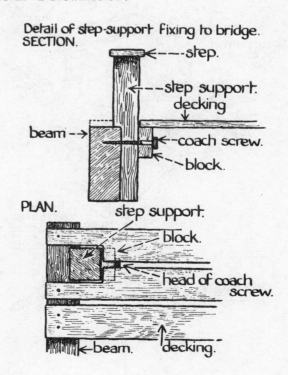

Detail of step-support fixing to bridge.
SECTION.

step
step support.
decking
beam
coach screw.
block.

PLAN.

step support.
block.
head of coach screw.
beam.
decking.

Single beam bridges usually only require a single bar to keep stock off. Some situations may require a watergate underneath to prevent stock going up or downstream.

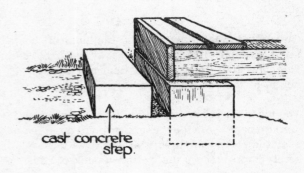

cast concrete step.

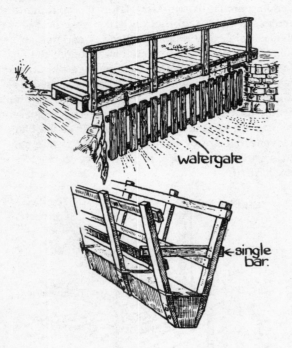

watergate

single bar.

The following two examples show how the appearance of a bridge can be spoiled by poorly-constructed steps, especially where several different types of materials are used.

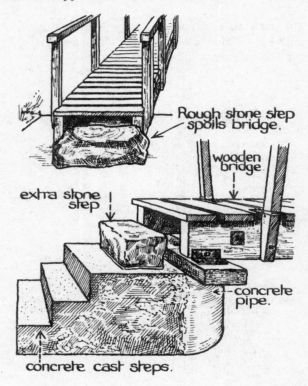

Rough stone step spoils bridge.

extra stone step

wooden bridge.

concrete pipe.

concrete cast steps.

STEPS

Construct steps with care so they do not spoil the appearance of the bridge. Two methods are shown below, using stone or cast concrete.

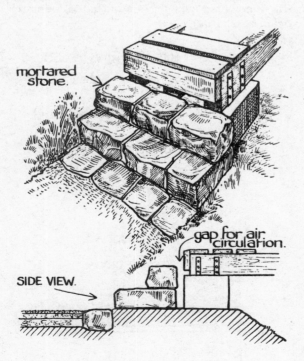

mortared stone.

gap for air circulation.

SIDE VIEW.

Design and construct carefully the approach to the bridge. Often this acts as a gathering point where walkers wait to cross or admire the view, and an area may need to be surfaced for this purpose.

SCOUR PROTECTION

This may be needed for a metre or so upstream and downstream of the abutment on fast-flowing streams. The downstream side can become scoured as water eddies around the abutment.

Boulders

These are simply piled up on either side of the abutment. This is the simplest but least durable method.

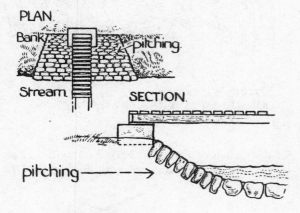

Stone pitching

Set the stones with their long axis into the bank. Make sure the profile is smooth so that the water flows unimpeded.

Revetments

These are described more fully in Chapter 11. Listed below are products designed for bank stabilisation, and which could be used to protect banks around bridge abutments and end seats.

Monoslab E (p182 9.1). This is a concrete slab with pockets which can either be filled with earth in which plants establish, or filled with aggregate. A slab is 400mm x 600mm x 110mm and weighs 42.5kg.

Gabions (p182 9.2). These are plastic coated wire mesh cages, which are filled with rock from the site. They are not attractive initially, but will become hidden with vegetation in suitable habitats after a year or two.

gabions - - - - →

Dytap interlocking blocks (p182 9.3), which are available in various sizes. Type B weighs 26kg, and is laid at the rate of 6.6 blocks to the square metre.

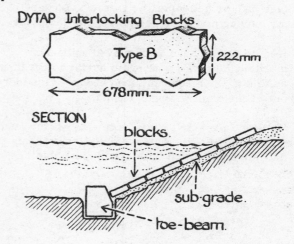

Wyretex no. 7 or 8 (p182 8.2). Rake the slope, then lay the Wyretex and pin at 500mm intervals. Cover with a 50mm layer of soil, tamp down, and seed.

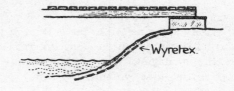

Procedures

SAFETY

Bridgebuilding is potentially dangerous, because of having to handle heavy and bulky main beams. The safest way is to use lifting plant such as a lorry-mounted crane, but many sites will not have suitable access. Below are some methods of moving and positioning main beams without using machinery.

The moving and positioning of main beams requires plenty of people, but with one person only to give the orders so there is no confusion. Once the beams are in position, the construction of the decking and handrails is best done by no more than three people. If too many people try to work at the same time on an unfinished bridge they will get in one another's way, with potential danger if there is a drop to the stream bed. Other volunteers can be involved in preparation of decking and handrails, or in clearing and surfacing the bridge approaches.

CARRYING MATERIALS

This can be the hardest part of constructing a bridge. Special transport such as an Argocat (see p45) or helicopter may have to be used for moving main beams into remote locations.

If main beams have to be moved without a machine, always try to avoid having to lift them, as this is when accidents can occur. Consider one of the other methods first.

'Dollies' are small carts with wide wheels, used on building sites. A 'bomb dolly' can take up to half a tonne. Local builders may be able to lend one.

Timber skids can be nailed to the bottom of the beam so that it can be pulled along the ground without being damaged. Try and use horse-power, or failing that, volunteer-power.

Timber skid nailed to beam.

If the main beam has to be lifted, use carrying beams which can be securely gripped. Always tie the main beam to the carrying beams so it cannot move. Fix blocks to the bottom of the beams so that the load can be put down without fingers being trapped.

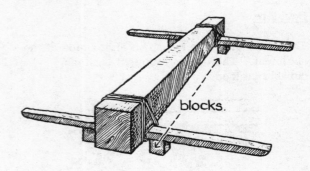

blocks.

Try and take some of the weight on shoulder slings. These can be made of old seat belts from scrap yards.

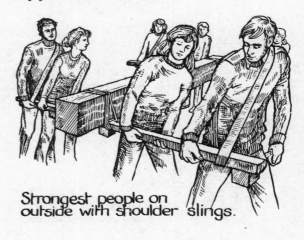

Strongest people on outside with shoulder slings.

Railway sleepers can be carried using metal pins pushed through existing holes in the sleeper.

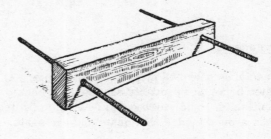

POSITIONING THE MAIN BEAMS

The procedure must be carefully planned. For shallow streams and ditches, it should be possible to manhandle the beam into position. The remaining beam or beams can then be slid along the one already in place. Beams which are too heavy to manhandle will have to be brought to the site by machine, and must be lifted into place by a lorry-mounted crane or excavator arm.

To manhandle beams across deep gaps, use one of the following methods:

a Place one beam alongside the abutment, and using a heavy weight to counterbalance, slide it out across the gap. The next beam can then be pushed along it and into place.

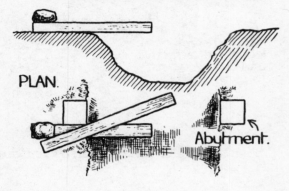

b Pull the beam across the gap, using a stout tree or post to belay the pulley. If neither are available, make an anchor of metal pins lashed together with rope.

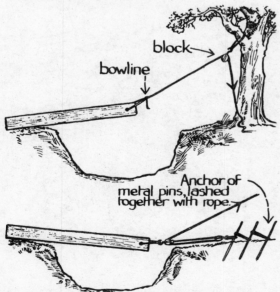

Rope must be at least three times length of gap.

c Use shear legs and spars. This should be done by an expert.

ORDER OF WORK

Design

This summarises the stages in designing a bridge.

1 Gain consents.

2 Survey the site.

3 Note materials available, resources, access and expected use.

4 Choose bridge type and width.

5 Consult the table of beam sizes (p106).

6 Design abutment or end seat, and fixing for main beams.

7 Calculate handrail post spacing and handrail size (p107).

8 Calculate spacing and number of deckboards.

9 Make a complete materials and tools list.

10 Decide on method of getting materials on site.

Construction

This summarises the stages in constructing a bridge.

1 Construct abutment or end seat and scour protection. Allow time for concrete to set.

2 Place damp-proof membrane on abutment/end seat.

3 Position main beams. Wedge and secure struts.

4 Lay bituminous felt on beams.

5 Attach deckboards, except for those between the handrail posts.

6 Fix handrail posts and tighten steel ties.

7 Attach remaining deckboards.

8 Attach handrails.

9 Build steps and/or stile.

10 Surface approach path.

10 Steps

Building steps is one of the most difficult types of footpath construction, and often fails due to wrong choice of line, insecure construction, or lack of drainage. Only build steps if there is no other way around the problem. The following are guidelines to consider when deciding whether or not to build steps:

a   On existing routes, is the slope so badly gullied or eroded that steps are required for the prevention of further damage?

b   Is there danger to path users because of an eroded or slippery slope? Danger may be acceptable in some locations, such as on mountainsides, whereas a path in a country park should cater for the less agile walker.

c   Consider whether there is any provision for maintenance. Well-built stone steps should be maintenance-free, but steps with wooden risers require frequent attention and it may be better not to build if such provision for maintenance cannot be made.

d   Are there alternative ways which visitors can use to get up and down the slope? What are the chances of walkers keeping to the steps?

e   Try to anticipate where steps will be needed on new paths, instead of waiting to see where erosion occurs. It is easier to get steps into use if they are part of the original design. Putting them in when the need arises will involve the extra work of repairing damage and changing patterns of use.

Designing and Estimating

LINE AND LOCATION

a   Avoid straight lines. Long straight flights of steps are an intimidating prospect for the walker, and look out of place in rural settings. Break up long flights with bends and ramps.

Straight steps are also more difficult to drain effectively, as water tends to collect behind each riser, or run down the sides of the steps, causing erosion.

Water gullies down sides, and collects behind steps,

BUT-

-curved steps look more attractive, and appear easier to climb, and-

- water can be taken off at sides.

Cliff-top paths may well have no alternative to the direct line. In this case, the steps should be built as close as possible to a staircase, with even width and gradient, landings where possible, and a handrail.

b   When choosing the line of the steps, look at the site from both above and below. Walkers going down are more likely to take short cuts, and it is this that causes most of the damage to slopes. Make sure that the line can be clearly seen from above, and that any possible short cuts are blocked or disguised.

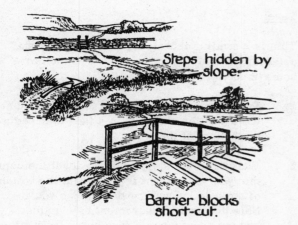

Steps hidden by slope.

Barrier blocks short-cut.

c   Avoid building sidehill steps, especially across unstable slopes. They are difficult to construct, and need abutments above and below. The treads tend to drain too quickly, causing erosion on the lower side. If possible, build on a steep oblique line, which is easier to drain.

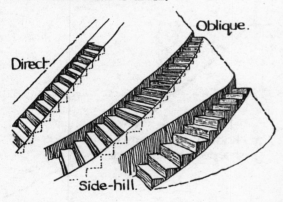

Direct.

Oblique.

Side-hill.

DIMENSIONS

Risers should be 200mm high, and at least 1000mm wide for two way use. Variations in the slope are accommodated by altering the depth of the tread.

The dimensions of stone steps will be dictated by the stone available, but must be at least 600mm wide, which only allows single-width use. Steps for two-way use can be built of several large stones, to make a step at least 1000mm wide.

600mm. Min. one-way.

1,000mm. Min. two-way.

Wooden steps should be 1200mm wide to allow comfortable two-way use. Wide steps look more

attractive and less steep than narrow steps. Both for these reasons, and because they give more room for traffic flow, wide steps are likely to be followed by walkers. The timbers themselves may be wider than the finished step width, depending on the method of construction.

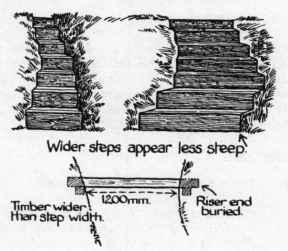

Wider steps appear less steep.

Timber wider than step width.

1200mm.

Riser end buried.

The depth of the tread depends on the gradient of the slope.

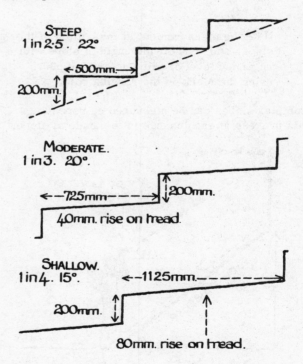

STEEP.
1 in 2·5 . 22°

500mm.

200mm.

MODERATE.
1 in 3. 20°.

725mm.

200mm.

40mm. rise on tread.

SHALLOW.
1 in 4. 15°.

1125mm.

200mm.

80mm. rise on tread.

500mm is the minimum depth of tread and should only be used on 'staircase' type flights, and should be broken where possible with a landing. Most slopes have a changing gradient which can be climbed with flights of one or two pace steps, divided by ramps. It is not possible to maintain a constant depth and tread on a variable slope, but try to keep the same depth of tread in any one flight. As shown, treads should always be built with a rise from front to back. This is not only very important for drainage (see below), but saves on the total number of steps needed, and forms steps that are comfortable to walk up and down as they do not break the rhythm of walking.

ESTIMATING MATERIALS

It is important to survey the slope in some way in order to design the flights and landings or ramps, and estimate the amount of materials needed. If surveying equipment and expertise is not available, either of the following methods can be used. Both can be done by one person on their own.

Clinometer

To measure a simple slope of even gradient:

1   Put a ranging pole or cane, with an easily visible mark on it at your eye height, in the ground at the top of the slope. Attach the end of a tape to the bottom of the pole.

2   Walk down the slope, unreeling the tape as you go.

3   Note the measurement at the bottom of the slope, and read off the angle of slope with an Abney level or clinometer (see p20) using the mark on the ranging pole.

Complex slopes can be measured by repeating this process at each change in the angle of slope.

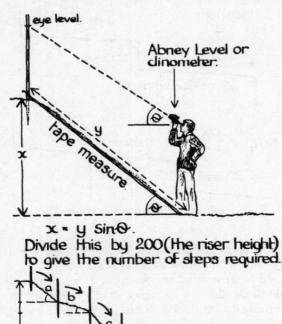

$x = y \sin\theta$.
Divide this by 200 (the riser height) to give the number of steps required.

Complex slope.

Level

This is a similar technique, but measures the height instead of the angle. Use a sextant or a home-made level (see p98).

1   Put a ranging pole or cane in the ground at the top of the slope, with a tape attached to the bottom of the pole.

2   Back down the slope until the level indicates that your eye height is level with the bottom of the pole, unreeling the tape as you go. Note the tape measurement to your feet, and put down a temporary mark (eg the level or your notebook) at that spot.

3   Retrieve the ranging pole, stick it in the ground at the marked spot, and repeat the process.

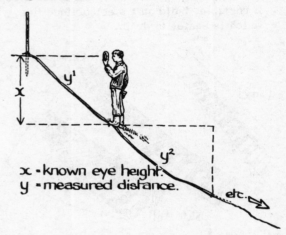

x = known eye height.
y = measured distance.

This information can be drawn directly onto graph paper. At a suitable scale, draw in the vertical line x. Place the ruler through point A, and swing it until y' meets the line CB.

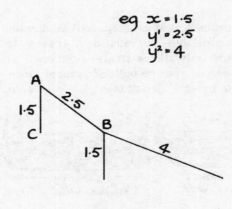

eg x = 1.5
y' = 2.5
y² = 4

Timber Steps

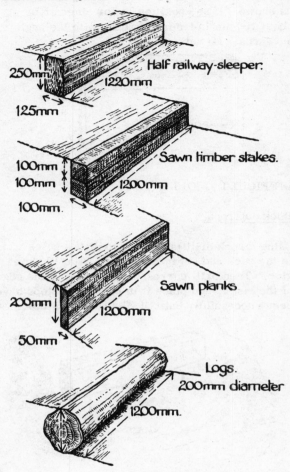

Half railway-sleeper:
250mm
1220mm
125mm

Sawn timber stakes.
100mm
100mm
1200mm
100mm.

Sawn planks
200mm
1200mm
50mm

Logs.
200mm diameter
1200mm.

STOBS

These secure the riser to the ground, and can be square or round timber stakes, metal pins or angle iron. Avoid using more than two stobs for each riser, and always choose to use two longer rather than three shorter stobs. It is difficult to knock the third stob in exactly in line so that the stress is placed equally on all three stobs.

If two stobs are not sufficient to hold the riser without it bending, then the riser is not strong enough. If unsuitable timber has been supplied, or if a repair job is being done, put the middle stob on the inside and nail.

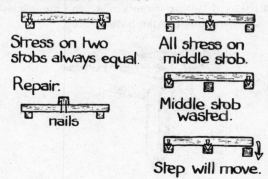

Stress on two stobs always equal.

All stress on middle stob.

Repair.
nails

Middle stob wasted.

Step will move.

Normally the stob should be nailed to the riser, as this makes the construction much stronger.

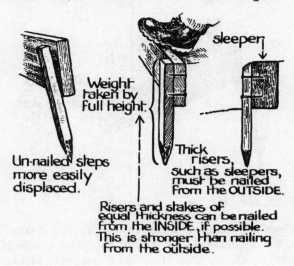

Un-nailed steps more easily displaced.

Weight taken by full height.

sleeper

Thick risers, such as sleepers, must be nailed from the OUTSIDE.

Risers and stakes of equal thickness can be nailed from the INSIDE, if possible. This is stronger than nailing from the outside.

Square stakes

Use 50mm x 50mm timber. The length will depend on the type of ground, but will normally be about 450mm. In ground where stakes can be knocked in easily, the step can be made stronger by cutting a 15mm notch, as shown. Nail using galvanised nails, and weather the top of the stob. Do not cut notches for stakes in stony ground, as it will be difficult to get the stakes in exactly in line with the notches.

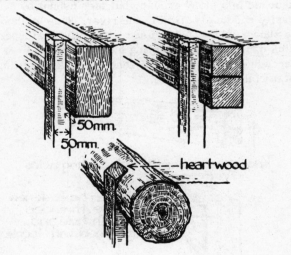

50mm.
50mm.
heartwood

Always place the stake with the heartwood against the riser, as shown, as it is more durable than the sapwood.

The main disadvantage of square stakes is that it is difficult to knock them 'squarely' into stony ground. Often they twist slightly as they go in, leaving a gap which prevents a strong, close joint being made. Try starting the hole by working a crowbar across the 'diagonals' of the hole, and use a stob holder to hold the stob firmly.

117

Rounding the point of the stake reduces the tendency to twist.

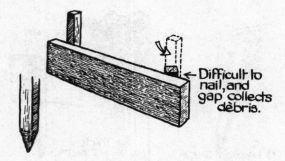

Round stakes

These are easier to knock in than square stakes, as the alignment is not so critical. Use stakes of about 75mm top diameter. Alternatively, larger stakes can be split into half or quarter rounds, which are easier to nail than round stakes.

Metal pins

These can be made of reinforcing rod or any other suitable material to be found at scrap merchants or demolition sites. Road pins or bunting pins used by local councils are another possible source.

Metal pins have the advantage that they can be knocked into stony ground, but they cannot be easily and neatly fixed to the riser. One method is shown below. When used with railway sleepers, pins can double as carrying handles, pushed through existing holes, to ease the task of getting materials up to the site.

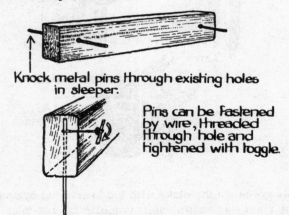

Knock metal pins through existing holes in sleeper.

Pins can be fastened by wire, threaded through hole and tightened with toggle.

Angle iron

This makes neat and inconspicuous stobs for sleeper or sawn timber risers, and in most situations will become hidden as vegetation covers the sides of the steps. Nail with galvanised nails. In stony ground, it may be difficult to accurately

position the angle iron at the corner of the riser, as shown. If so, position as for other stobs, about 100mm in from each end, with the angle outwards. Do not nail.

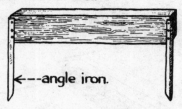

angle iron.

DIFFICULT GROUND

Rock outcrops

Using a rock drill (see p45), drill two holes in the rock and set metal pins in with 'Rockite' glue. Then drill corresponding holes in the edge of the riser, and place it in position. Steps have been successfully built at Glencoe using this method.

Stony ground

Excavate as deep a hole as possible using a crow-bar, and concrete the stob in position.

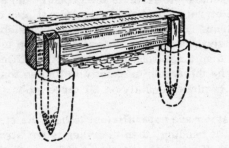

Small patches of difficult ground can be climbed by buttressing the steps. This can only be used for a flight of two or three steps, and must have very secure stobs for the bottom step.

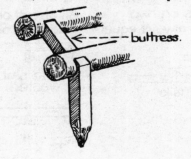

buttress.

Drainage and Surfacing

A common problem with wooden steps is that the tread compacts with use and water collects behind the riser. Trampling turns the surface to mud, which is carried away on boots or washed away by rain. The steps are abandoned as they become uncomfortable to use, and a path forms alongside.

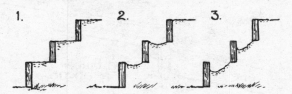

This problem can be prevented by proper surfacing and drainage:

a   The tread should be formed of free-draining material which will not be carried away on boots. Do not use clay or organic soil.

b   Prevent 'waterfall' flows down the steps by constructing a cross-fall or camber to shed water off to the side. On steep slopes and impermeable ground, install a French drain down the side.

Treads should always slope from back to front (see p115) to allow water to drain off. Do not sink the bottom of a riser below the top of the previous riser in an attempt to prevent the tread eroding, as this only makes worse the effect illustrated above. To reduce the waterfall effect down the face of the steps, the tread should have a cross-fall or camber to shed water to the side. A camber will not be as durable as a cross-fall, but should be used on steps taking a direct line up a slope.

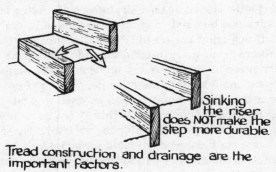

Sinking the riser does NOT make the step more durable.

Tread construction and drainage are the important factors.

On steps where the tread mainly comprises 'fill' material, the tread can be built up in a similar way to a surfaced path (see p79). Local material from borrow pits, scree or stream beds should be suitable. Grade the material as shown, and put in a cross-fall or camber to direct the water away from the step.

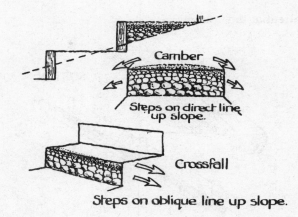

Camber

Steps on direct line up slope.

Crossfall

Steps on oblique line up slope.

Steps where the tread is partly comprised of fill can have a 'toe' or French drain of free draining material to take water from the tread.

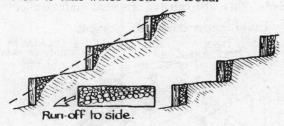

Run-off to side.

Steps cut into steep slopes of clay need a surface dressing to prevent the treads becoming muddy and eroded. Crushed stone of about 20mm to dust, well trodden down, is suitable. This will need renewing periodically.

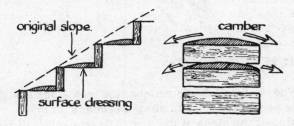

original slope.

camber

surface dressing

Side drains

On moderate clay or peat slopes, an open ditch can be dug. This should be about 300mm from the edge of the steps, to prevent them being undermined. Water must be able to flow freely into the ditch, with no possibility of it being diverted back onto the flight of steps. Make the gradient of the ditch as smooth as possible, and on long flights, lead it off at intervals to soak into the slope at a distance from the steps.

On steep slopes or those likely to slump, a French drain with a perforated plastic pipe should be constructed. This can be immediately adjacent to the step. Alternatively, stone-lined drains (see p73) can be built. Remember to run the side drain into a ditch or soakaway, so the water does not merely collect at the bottom.

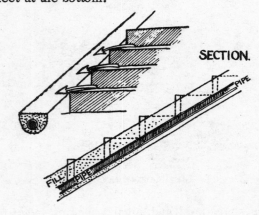

SECTION.

On very steep steps, the following method can be used. This will require frequent maintenance.

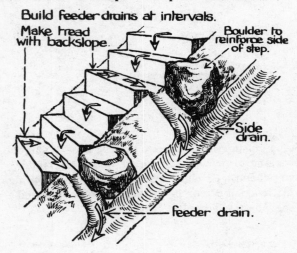

Build feeder drains at intervals.
Make tread with backslope.
Boulder to reinforce side of step.
Side drain.
feeder drain.

REVETTED STEPS

Where steps take an oblique line up a slope it may be necessary to build simple revetments to protect the sides of the step. Revetments will be essential where steps have to climb an unstable slope.

Sawn timber

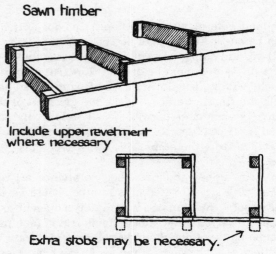

Include upper revetment where necessary

Extra stobs may be necessary.

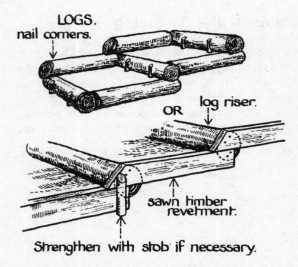

LOGS.
nail corners.
OR log riser.
sawn timber revetment.
Strengthen with stob if necessary.

Install a French drain on the lower side.

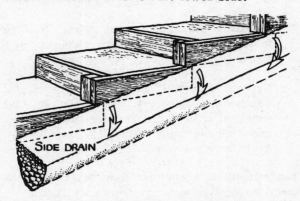

SIDE DRAIN

Revetted steps should only be built where no better line is available, as they require a lot of timber and can look very obtrusive. See also Chapter 11 for other types of revetments which can be built in association with steps.

SUNKEN STEPS

On gentle slopes of free-draining soils extra long risers can be used, and the ends buried in the spoil dug from the tread. This discourages walkers from going around the side, and looks attractive as the stobs are hidden and the steps appear moulded into the hillside. Any turf cut from the line of the steps can be used to protect the shoulder of the slope. This is not recommended on steep impervious slopes, or where run-off is high, as the unconsolidated material will be simply washed away as the line of the steps acts as a watercourse.

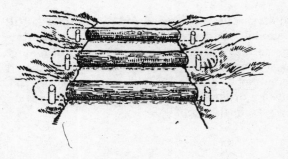

In clay soils it may be possible to cut slots to protect the ends of the risers, leaving a shoulder of undisturbed soil. Alternatively, spoil can be piled up against the ends of the risers, but will be washed away if there is much surface run-off.

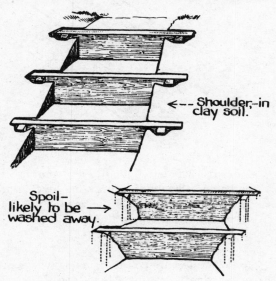

← Shoulder, in clay soil.

Spoil-
likely to be →
washed away.

NON-SLIP TREATMENTS

Log steps can get very slippery, especially in woodlands where humid conditions favour algal growth. Although the bark may seem to improve the grip, it should always be removed before the step is built, as it can itself become slippery and speeds the decay of the timber. Creosoting helps to reduce algal growth, and can be repeated at intervals as necessary.

The simplest method of improving the grip is to roughen the top of the step using a billhook. Alternatively, a recess can be cut in the top of the log using an axe or chain-saw. Do this after the step has been fitted, but before it is backfilled or surfaced.

The method shown below has successfully withstood several years' heavy use at the Birks of Aberfeldy, Tayside. It is not too obtrusive, as the wire soon loses its silvery appearance.

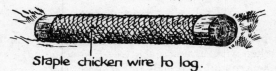

Staple chicken wire to log.

Organising the Work

On continuous flights of steps, or flights where landings and turns must be very accurate, work can proceed only from the bottom upwards. Usually, flights are divided by ramps and so separate groups can work on the different flights. The job can be divided as follows:

1   Carrying materials, including surfacing.

2   Cutting the steps, fitting the risers and knocking in the stobs.

3   Cutting the side drain.

4   Nailing the stobs, weathering the top of the stob and treating with creosote. Roughening or fixing chicken wire to log steps, as necessary.

5   Backfilling and surfacing the tread.

Note the following points:

a   Place the riser and secure it before proceeding to the next step. In clay soils the full height can be excavated before the riser is fitted. In less cohesive soils, the riser will have to be sunk in part of its height, and then material cut from the front and put behind the riser.

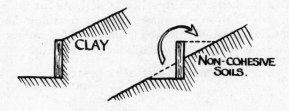

CLAY

NON-COHESIVE SOILS.

b   Steps cut into steep slopes will leave an excess of spoil. Stony material can be stockpiled nearby to repair the treads as they become worn. Re-use turf on the shoulder of the step, or elsewhere on the path. Soil should be scattered away from the line of the steps.

c   Use the full height of the 200mm riser, so that the slope is climbed with the minimum of steps. Any erosion at the bottom of the riser will be avoided by proper drainage and surfacing.

d   Make sure that the riser is absolutely firm in the ground. If it rocks at all, it will loosen with use.

Timber Ladders

These are used to climb very steep or unstable slopes where steps cannot be built. This might include a steep slope down a river bank leading to a bridge, or a short section on a hillside where it avoids having to build a long hairpin detour.

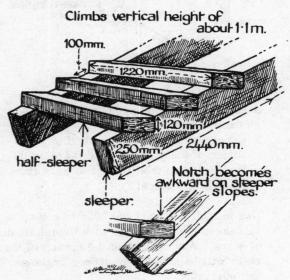

Railway sleeper

This method of construction is suitable for slopes of about 1 in 2. On slopes steeper than this the notches become awkward to cut, and the alternative method using bearers is easier to build.

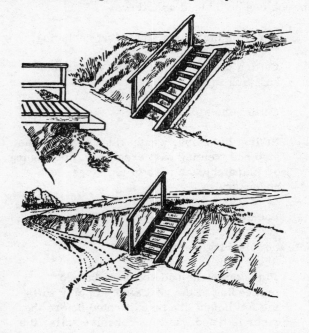

Climbs vertical height of about 1·1m.

100mm.

1220mm.

120mm

250mm 2440mm.

half-sleeper

sleeper.

Notch becomes awkward on steeper slopes.

1   Excavate the line of the ladder to about 250mm deep, and set the two stringers in place, with their outside edges parallel and a half sleeper length apart.

2   Using a levelling board and spirit level, set and mark the position of the first step, and then mark in the notch to the required dimension.

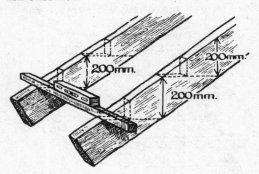

200mm.

200mm.

200mm.

3   Measure 200mm vertically up from the notch on either side, check the level, and mark in the position of the second step. Continue to the top, which should total seven steps. Cut the notches.

4   Nail the steps in position, using wedges if necessary to get them level and solid. The weight of the structure will hold it in place.

Timber ladder with bearers

This can be built on slopes of up to 1 in 1.2. The angle and length of the slope should be measured accurately using one of the methods described on page 116 in order to estimate the materials required.

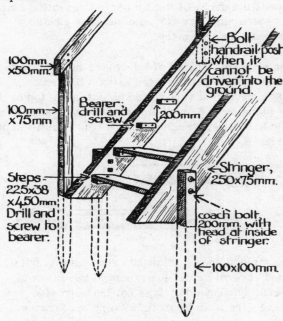

100mm x50mm.

100mm x75mm

Bearer; drill and screw.

Bolt handrail post when it cannot be driven into the ground.

200mm

Steps 225x38 x450mm Drill and screw to bearer.

Stringer, 250x75mm.

coach bolt 200mm with head at inside of stringer.

100x100mm.

To prevent the ladder becoming slippery, staple chicken wire around each tread, making sure there are no loose ends underneath which could catch on footwear.

Stone Steps

There are three basic ways of building stone steps without the use of mortar, and all three require stones that are large enough not to move underfoot. Use the largest available stone that can be moved into position.

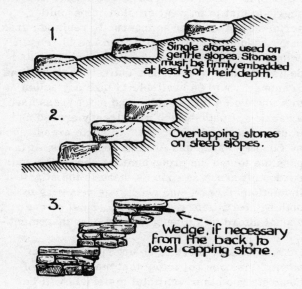

1. Single stones used on gentle slopes. Stones must be firmly embedded at least ⅓ of their depth.

2. Overlapping stones on steep slopes.

3. Wedge, if necessary from the back, to level capping stone.

Although the capping stone does not have to be as high as in the other two types of steps, it must be large and heavy enough not to move underfoot, and to hold the smaller stones in position. Building these steps successfully requires skill and practice.

Stone and concrete

If no large stones are available, concrete or mortar can be used to hold the stones in position. Only do this as a last resort, and do not use at all in the mountains or on slopes that are unstable. Shattered concrete is a far worse eyesore than an eroded path.

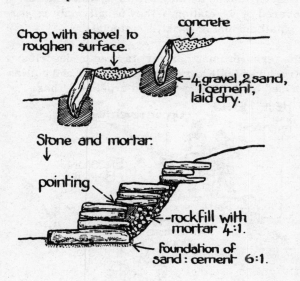

Chop with shovel to roughen surface.

concrete

4 gravel, 2 sand, 1 cement, laid dry.

Stone and mortar.

pointing

rockfill with mortar 4:1.

foundation of sand : cement 6:1.

11 Erosion Control and Vegetation Restoration

Erosion is a serious problem on many popular paths and areas where the public have unrestricted access. Excessive trampling destroys the vegetation, so exposing the soil to rain splash and overland flow, and causing a great increase in the rate of erosion. The steeper the slope, the greater is the problem. Damaged landscapes range from mountain tops to peat moorlands, chalk downs, lowland heaths and coastlands.

There are various ways of tackling the problem:

1   Restricting types or amount of use by entrance fees, bye-laws, and other management techniques. This is outside the scope of this book, but see Chapter 3 for notes on use and carrying capacity of paths.

2   Spreading use over a wider area to lessen the impact on one particular path or location. On the large scale, this may be part of a management scheme for an extensive area of countryside, such as a National Park. On the small scale, this may require simply the removal of fences enclosing a path to allow use to spread. On chalk downland and ancient monuments particularly, the aim is often to encourage an even use all over the site by removing 'targets' and desire lines, or in other words, allowing access without the development of paths. This is discussed in Chapter 4.

3   Draining and improving the path, or altering its line. See Chapters 3 - 10.

4   Restoring damaged areas by constructing revetments and erosion barriers, filling, grading, reseeding and transplanting. These techniques are considered below. Successful restoration requires keeping people off the area being restored, and is a complementary process to improving the path. Techniques to encourage users to keep to the path are discussed in Chapter 4.

THE MOMENT TO ACT

In many places the moment for a 'stitch in time' passed by many years ago, and the problems are now enormous. This is no reason for not attempting something. As shown at Arnside Knott and Nab Scar in Cumbria (see p130), it is possible, with lots of enthusiasm and many willing hands, to succeed against the apparently impossible.

Studies have shown that soil structure is damaged by trampling while vegetation is still maintaining

itself, and by the time there is visual evidence of declining plant cover, the critical period in which erosion is initiated is already past (Quinn, Morgan and Smith, 1980).

This presents a dilemma to anyone responsible for deciding when action should be taken, as it suggests that to prevent erosion one should surface or strengthen an apparently resistant grass path, assuming that levels of use are to remain constant. This would obviously be an unpopular decision. In practice, it is unlikely that the time or money would be available to take any action at this stage on a large scale, and most areas have a backlog of paths which are already eroded and in urgent need of treatment. In some areas, such as Country Parks and Ancient Monuments, it is possible to use the groundsman's approach, and cordon off areas as soon as there is the slightest indication of wear, and encourage recovery by spiking, fertilising and seeding. Most places cannot afford this intensive level of management.

On many sites, there is an urgent need to act now. This does not imply that massive use of concrete and other artificial materials will be necessary, or that all susceptible paths should be 'engineered'. Changes made by volunteer action will be labour intensive, mostly small-scale and use mainly local materials.

Revetments

In this book the term 'revetments' is used to describe structures which retain slopes below or above paths, and prevent the path either eroding from below, or being covered with material slumping from above. Revetments should have an attractive finish as they will be visible unless covered by vegetation. They usually only require a small amount of backfill material.

The term 'erosion barrier' is used to describe structures built to repair bare slopes and gullies caused by trampling-induced erosion. These

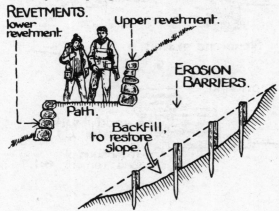

barriers are usually buried, and therefore their appearance is not important. Erosion barriers require a large amount of backfill material in order to restore the slope to its original profile. Most designs can be used for either purpose, as their function is similar.

LOG REVETMENTS

These are appropriate on wooded slopes where there is abundant timber. Vertical revetments up to 1.5 metres high can be built, provided that posts can be knocked in securely. Buttressing may be needed. A similar design can be used for erosion barriers.

SECTION.

Nail from inside if possible.

Overlap logs on long straight sections.

BUTTRESSED LOG REVETMENT for lower side of path.

buttress.

Notch post to give secure fixing for buttress.

BRACED TIMBER REVETMENTS

This technique uses a 'skin' of sawn timber tied by wire to bracing stakes buried in the bank.

The following design has a skin of posts and rails with chain-link fencing, enclosing a rubble wall. It can be used to support an unstable slope above a path, where space is limited and the path will be 'lost' if the slope collapses. The chain-link fencing is not attractive, but becomes less obtrusive as vegetation grows through. This can

be encouraged by filling between the fencing and the wall with turves. This design is suitable for erosion barriers.

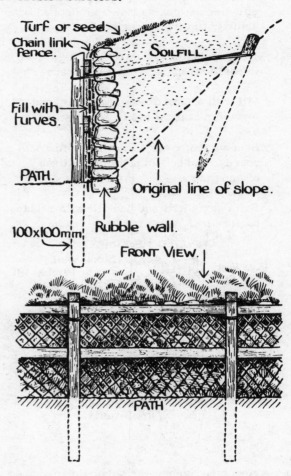

Turf or seed.
Chain link fence.
SOILFILL.
Fill with turves.
PATH.
100x100mm
Rubble wall.
Original line of slope.
FRONT VIEW.
PATH

The very strong construction described below should only be necessary for revetting banks of streams or rivers along which paths pass. The procedure for building is as follows:

1   Clear debris and loose rock from the line of the revetment.

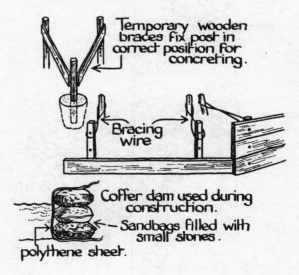

Temporary wooden braces fix post in correct position for concreting.

Bracing wire

Coffer dam used during construction.
Sandbags filled with small stones.
polythene sheet.

2 Construct coffer dam to keep as much water as possible out of the work site. It is difficult to make it watertight, but is effective as long as water is stopped from actually flowing through the site.

3 Excavate holes and carefully align the posts in position, using temporary braces.

4 Fill with 3:1 concrete (see p186), and leave for at least two days to set.

5 Remove temporary braces. Drill posts and boards, and bolt in position. Attach bracing wire.

6 Build stone wall and backfill with rubble. Surface path as necessary.

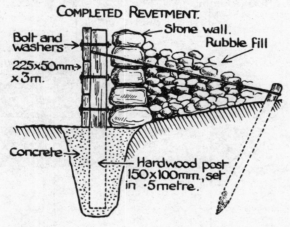

COMPLETED REVETMENT.

Bolt and washers
Stone wall.
Rubble fill
225x50mm. x 3m.
Concrete
Hardwood post 150x100mm. set in ·5 metre.

In the revetment below, the chain-link or sheep fencing is continued over the top to enclose the

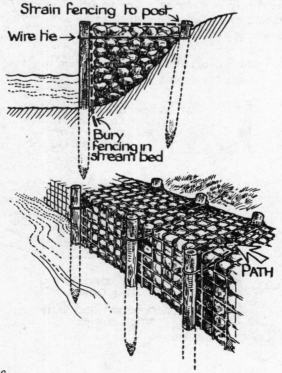

Strain fencing to post
Wire tie
Bury fencing in stream bed
PATH

rubble fill, in a similar way to a gabion (see p128). It can be used to build a path alongside a river which frequently floods, and provided the rubble is larger than the mesh size, should not get washed away. It is not very attractive, and the path surface is rough, but it does provide a cheap solution that uses a minimum of purchased materials.

DRY STONE REVETMENTS

The technique of building a dry stone revetment is similar to that of building a dry stone wall. The basic procedure is as follows:

1 Cut back the bank as necessary, to the sharpest angle possible without it slumping. Cut back enough to allow for the width of the stone plus some room to work. In order to keep the work site tidy, and prevent slumping of slopes, only prepare as much as you can revet that day.

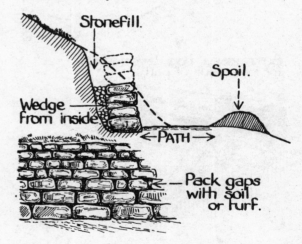

Stonefill.
Spoil.
Wedge from inside
←PATH→
Pack gaps with soil or turf.

2 Pile the material neatly along the outer edge of the path, leaving a clear path for walkers so they do not have to spread down the slope or trample the spoil material.

3 Excavate a trench about 150mm deep for the foundations.

4 Starting with the largest stone available, build up the revetment in horizontal courses, placing the stones level or sloping down towards the bank, with their long axis going into the bank. If wedges are necessary, place them from the inside. Pack behind each course with stone filling.

5 Some wallers like to pack turf and soil between the stones to encourage growth of vegetation which binds and camouflages the revetment.

6   There are various ways of finishing the top of the revetment, depending on the situation, the stone available, and the local walling style. If suitable stones are available, they can be laid flat on the top of the revetment, with the ends embedded in the bank. They must be large enough that they cannot easily be dislodged. Alternatively, the top can simply be covered with turf.

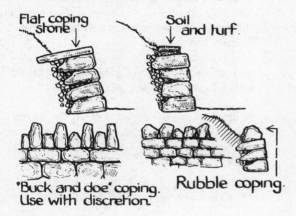

Low revetments are very likely to be used as seats, which may or may not be desirable. This use can be discouraged, for example on narrow paths, by either building the revetment too high to be comfortable, or by placing a rubble coping on the top. The 'buck and doe' style of coping should only be used where it is part of the local walling style, as it is too decorative for many rural situations.

The method of building revetments described above is suitable for revetments up to about one metre high, either above or below the path. If built below the path, top with turf or flat coping stones. Do not use rough coping stones in an effort to keep people away from the edge, as the stones will only get dislodged. The revetment must anyway be strong enough to withstand the weight of people standing near the edge. The revetment can be extended up into a low wall, if this is thought necessary for safety.

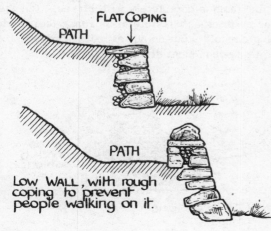

If necessary, build two or more revetments, each up to a maximum of one metre, rather than one very high revetment. Although massive slumps cannot be prevented, these revetments reduce most types of erosion on unstable banks.

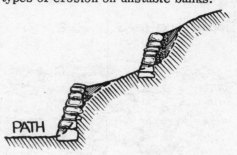

A quick method to protect the face of a slope is to simply build up stones against the slope, packing soil and turf between them to bind the surface. Like all revetments, this is merely slowing down the inevitable process of erosion, and cannot be guaranteed a long life. However, this technique has been used successfully in parts of Snowdonia to reduce erosion and encourage walkers to keep to paths on unstable slopes.

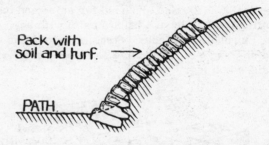

Steep revetments over one metre tall should be built as retaining walls. Two different designs are shown below. For further information on walling see 'Dry Stone Walling' (BTCV 1977).

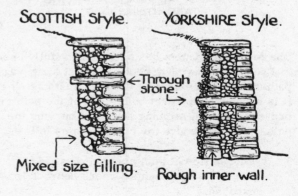

MORTARED REVETMENTS

Mortared stone revetments should be built along riverbanks or in any situation where extra strength is needed.

The design below can be used on the outer bends of rivers where erosion is greatest. The stepped design looks less obtrusive than a straight wall, and allows the water to flow more smoothly as levels rise.

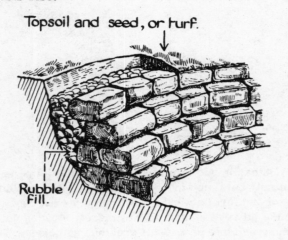

Topsoil and seed, or turf.

Rubble Fill.

GABIONS

These are oblong boxes of plastic covered wire mesh (p182 9.2), and are used for slope and bank stabilization. They are filled in situ with any stone or rubble available. Straining wires can be fixed at $\frac{1}{3}$ and $\frac{2}{3}$ of the volume as shown, to prevent the sides bulging.

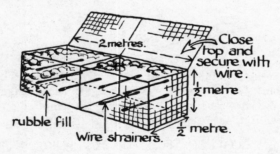

2 metres.

Close top and secure with wire.

$\frac{1}{2}$ metre

rubble fill

Wire strainers.

$\frac{1}{2}$ metre.

The following system has been successfully used at Fynnon Llugwy, Gwynedd, to revet a zig-zag path up an eroded scree. On such unstable slopes it is necessary to revet both sides of the path to prevent material slipping down and burying the path. The procedure for building is as follows:

1 Choose line of path along most stable sections, for example where bedrock is exposed.

2 Roughly level area for bottom gabion. Position gabion and fill. Position corresponding gabion on other side of the path.

3 Level area for next gabion, to overlap the one already filled. Continue up in this way, linking into bedrock or very large boulders

where possible. The line of gabions are less obtrusive from a distance if they are not too straight, so try and offset them a little to break up the line. This also makes the path more interesting to walk up, and less like a tunnel.

4 Lay boulders for steps between the lines of gabions, and if possible, surface with stone pitching (see p84). This greatly strengthens the path, and so prevents erosion undermining the gabions.

5 Cover the tops of the gabions with loose stone to disguise the wire.

6 Scatter stone over paths down the scree which you wish to close.

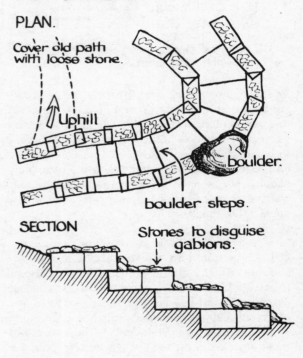

PLAN.

Cover old path with loose stone.

Uphill

boulder.

boulder steps.

SECTION

Stones to disguise gabions.

In areas with more vegetation, it is possible to hide the gabions completely with turves. Cut the turves as thick as possible to increase their water holding capacity, but do not expect success if there is a subsequent drought.

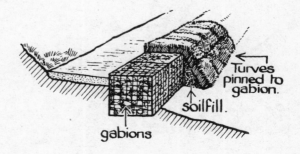

Turves pinned to gabion.

soilfill.

gabions

TYRES

It should be possible to obtain old tyres cheaply from tyre depots and garages.

1   Clean back the slope, and pile up the spoil for re-use.

2   Lay the bottom layer of tyres and fill with soil. Fill the gap behind with any available spoil. Compact with a roller or panner.

3   Continue placing and compacting the layers, and cap with a layer of concrete to prevent vandals dislodging the tyres.

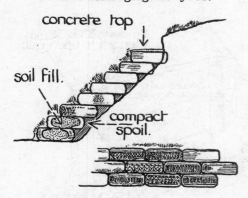

Depending on the situation, the tyres should be completely disguised with vegetation within a couple of years. This can be encouraged by including turves or roots of plants in the tyres, or by seeding. In shady situations, plant with ivy (Hedera helix) or periwinkle (Vinca minor).

Other methods of revetment include the use of concrete filled sandbags. These are best for bankside revetments, or for erosion barriers which will be mainly buried, as they are not attractive. Special revetment blocks are available which may be suitable for riverside revetments or for urban paths. See page 111.

Erosion Barriers

These barriers are built to reduce the erosion of slopes and gullies caused by trampling, in order that vegetation can re-establish. It is essential to keep walkers off the slope or gully, and therefore any scheme must include the building of a new path which takes a more resistant line, together with fences and publicity to encourage walkers to keep to the new path.

Gullies

These can occur on slopes of sand, chalk, or on much harder rocks in mountain areas.

The gullies considered here are those caused when trampling up and down a narrow zone destroys vegetation, exposing the thin soil and underlying rock to erosion by water. Erosion barriers should be placed to span the gully, as in the Nab Scar example on page 130.

In lowland areas of chalk or sand it is also possible to install drains to divert water from flowing down the gully (see p131). In upland areas, the high rainfall, high run-off, large catchment areas and the lengths of the gullies makes the construction of drains impractical.

Slopes

Eroded slopes occur where trampling is spread over a wide area, and flow of water is not concentrated in gullies, but extends over the whole slope. In some situations the slope will naturally stabilise and re-vegetate if a new path is provided and people keep off the slope, but in serious cases erosion barriers will be needed to stabilise the slope long enough for vegetation to take a hold. Other methods of stabilising slopes are described on page 134.

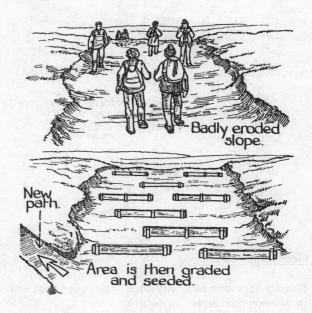

For either slopes or gullies, the erosion barriers are built in a similar way to revetments, but as they will eventually be partly or wholly buried, the appearance is not too important. The main requirement is for a simple and robust construction which can be easily positioned and secured. Working on steep and eroded slopes is very difficult, tiring and can be dangerous, and the work should be planned in such a way that the barriers can be positioned with the minimum of fuss.

The need to fill behind the barrier and grade the

129

slope depends on the situation. On very steep and rocky slopes it is impractical to do anything more than let the barriers fill as material moves down the slope. On Nab Scar (see below), the eroded gully was left open to walkers after the placement of the barriers, and the trampling helped to push material down behind the barriers and grade the slope. The gully was then closed to walkers. Where the gully reaches to the top of the slope and there is no way that material can naturally fill it, the barriers must be back filled and the slope graded by hand or machine. In other places grading may be necessary to speed the restoration process, or for cosmetic reasons.

The following examples show different methods of building erosion barriers:

Arnside Knott

These barriers were constructed on a rapidly eroding limestone slope, and have successfully stabilised the slope so that vegetation is now re-establishing.

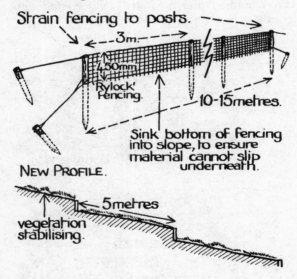

Greenside Mines, Cumbria

Simple barriers of 1.5m fencing stakes, knocked in at approximately 5m intervals on an unconsolidated tip slope.

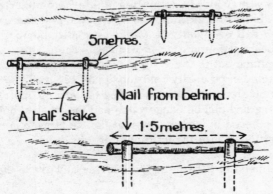

Note that the erosion in the two examples above was not initiated by trampling, but the problem is the same as on those slopes where over-use has caused the problem. It is usually best to start placing barriers from the bottom of the slope, so that any material dislodged during construction is trapped by a barrier below.

Nab Scar, Cumbria

A long, deep gully was sucessfully restored using the barriers shown below, and a new path built on a better line nearby. In spite of high rainfall and complete lack of topsoil, seeding has also been successful.

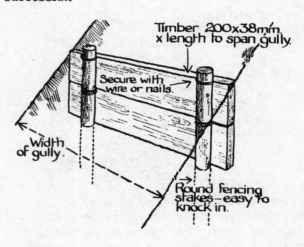

Cannock Chase, Staffordshire

This scheme involved the construction of a complicated series of erosion barriers and large-scale cut-off drains up a 100 metre long sandstone gully with a slope of 1 in 2.3. The barriers were made of sawn timber (green oak) and pine logs cut locally. They were secured by wooden stakes where possible, or by metal pins where the rock was resistant.

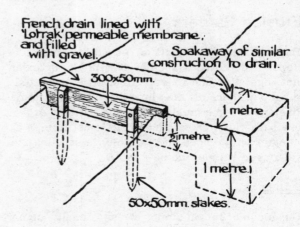

PLAN.

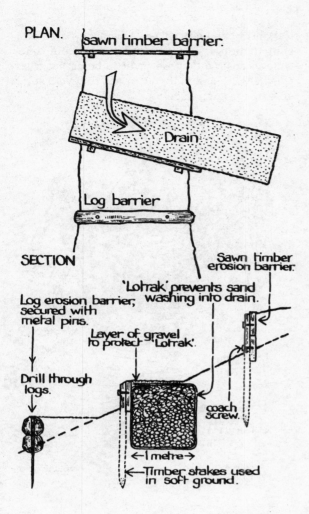

sawn timber barrier.

Drain

Log barrier

SECTION

'Lotrak' prevents sand washing into drain.

Sawn timber erosion barrier.

Log erosion barrier, secured with metal pins.

Layer of gravel to protect 'Lotrak'.

Drill through logs.

coach screw.

←1 metre→

←Timber stakes used in soft ground.

Badbury Rings, Dorset

Trials were done during 1981 to test different methods of restoring gullies on the hill fort of Badbury Rings. The gullies had been made over many years of heavy visitor use, which was restricted to narrow paths up the embankments because of scrub encroachment. The gullies were from one to five metres wide, up to one metre deep, and about five metres long from top to bottom of the chalk embankments. There was a 'toe' of eroded material at the bottom of each gully.

This work is more exacting than most erosion control work, because it is important to restore the slope to its original profile. The gullies will be monitored for several years to decide on the best method for this particular site.

Methods tried included the following:

a   Sawn timber with stakes. This is the standard design for an erosion barrier (see Nab Scar example), and is straightforward to construct. However, there is a tendency

for the backfill to slump behind each barrier, creating a stepped effect, which can then attract renewed trampling.

b   Turf dam. This is not as stable as a series of barriers, but is simple to construct, and avoids a stepped profile developing. It is probably most suitable for narrow gullies up to about one metre wide.

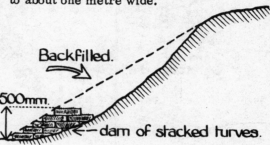

Backfilled.

500mm.

dam of stacked turves.

c   Chicken wire 'sock'. This would be a suitable method for shallow slopes, as it proved difficult to prevent the sock bulging at the bottom of the steep embankment.

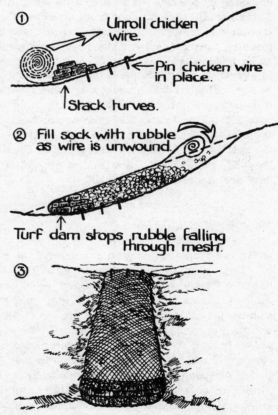

① Unroll chicken wire.

Pin chicken wire in place.

Stack turves.

② Fill sock with rubble as wire is unwound.

Turf dam stops rubble falling through mesh.

③

Filled sock pinned at sides. Cover with turves.

d   Ash wattles. These were made in situ from ash cut nearby, and proved time-consuming, but enjoyable, to make. Like the turf, they allow some flexibility which should prevent

131

the development of a stepped profile, and can
be made to fit irregularities in the gully.

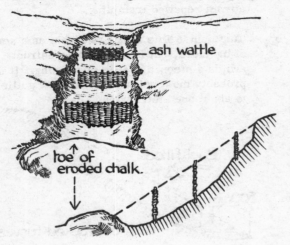

If there is access, machinery is useful for this
type of restoration, and as shown at Badbury Rings,
can be successfully combined with voluntary
manual labour. A Hymac excavator was used for
scraping up the 'toe' and backfilling the gullies.
Extra backfill was brought from elsewhere by
tractor and trailer. A crawler tractor was used
for turf-cutting nearby. The resulting turves were
about 225mm thick, which increased their chance
of survival, but made them more difficult to lay
neatly. The hardest manual work was positioning
the stakes as holes had to be made in the chalk
with a crowbar to about 200mm depth, before
the stakes could be knocked in. The surface of
the gullies was broken up with picks to provide a
'key' for the backfill. Methods of protecting the
turf are discussed on page 139.

MOVING MATERIALS

Gully and slope restoration can involve the moving
of a large amount of material for backfill. The
use of a Hymac as described above is the easiest
method if the site is suitable. It may be possible
to have the material dumped at the top of a slope,
and then move it by hand downhill. In the method
shown below, the backfill is shovelled onto a
'toboggan' of heavy tarpaulin, which is then
pulled at speed down the plastic chute by four
nimble volunteers. It is not possible to simply
push material down a polythene chute except on
excessively steep slopes (1 in 1.5 or steeper).

This method can be hazardous, because it must
be done at speed. If large groups of volunteers
are available, and particularly if they are young,
it may be safer, if less exciting, to simply move
the material by small amounts in buckets and
fertiliser sacks. There are sites where visiting

walkers have been asked to carry up a small
contribution. This needs checking from time to
time to ensure that material is being put where
you want it, and that you do not find yourself with
an embarassing surplus.

Vegetation Restoration

This can be done in three ways:

a   Allowing vegetation to return naturally, by
      seed spread or tillering from the surround-
      ing turf.

b   Sowing seed, either purchased or collected
      locally, of grasses and other plants.

c   Transplanting turves or clumps of vegetation
      dug from elsewhere. This may include
      nursery-grown trees and shrubs.

The success of any of these techniques depends on
keeping walkers off, at least for an initial period,
depending on the season and the rate of growth.

Natural recolonisation

In all but the most extreme locations of high
salinity, toxicity, or on moving scree, vegetation
should eventually return. The plants will only be
those occurring locally, and so a 'natural' plant
community should result. However, it may take
some years to achieve a balance, as in the first

few years the plant community will be dominated by pioneer or weedy (aggressive) species, usually those which are fast growing and can survive in extreme situations. In lowland areas for example, it is likely that thistles, docks and willow herb will arrive before the more desirable slower-growing grasses and herbs. After a few years, if the recovering patch is managed in the same way as the rest of the area, the weedy species will decline until the patch blends in unnoticed.

This natural pattern of recovery may be unsuitable for the following reasons:

a   The appearance of the pioneer plant community.

b   Growth may not be quick or thick enough to prevent renewed erosion of the area, particularly on steep slopes.

c   The rapid or even instant restoration of a scar or closed-off path may be a vital part of the scheme to persuade walkers to keep to the newly designated path. Walkers are more likely to respect the scheme if they can see that positive measures are being made to restore the eroded area.

Seeding

Seeding requires careful analysis and preparation if it is to be a success. This will include surveying the local flora, and analysing the soil or substrate to determine if any chemical treatments are necessary. Care must be taken that invasive species are not introduced which may upset the balance of species in surrounding plant communities. Seek advice if necessary from the local office of the Nature Conservancy Council.

The seed mix should usually be of native species and should include a fast growing 'nurse' species, and slow growing long lived species for permanent cover. As well as grasses, seeds of other herbaceous perennials, shrubs and trees can be sown.

After seeding the area must be fenced to exclude walkers, preferably for at least a year.

Turfing and transplanting

Like seeding, turfing will only be successful if the original reason for vegetation loss has been removed; usually by the provision of a more resistant path. Turfing has the obvious advantage that its effect is immediate, and it can transform the edges and surrounds to a new path so there is no question in the walker's mind of following any

route other than the path. If turfing is done with care, and the site is meticulously cleared up afterwards, it should be almost impossible to detect any signs of the work that has been done. This cosmetic effect can be vital in successfully altering visitors' patterns of use. Turfing is slow and sometimes heavy work, and the source of turves may be a limiting factor. Permission to cut turves must always be obtained from the landowner. Under the Conservation of Wild Creatures and Wild Plants Act 1975 it is illegal to dig up any wild plant without the landowner's permission.

Seeding

SOURCES OF SEED

Commercial suppliers

These are listed on page 182 (11.1 - 11.4).

Local seed

Seed can be gathered locally from grasses and other flowering plants. This is slow and pains-taking work, but it does mean that the seeds are of the same genotypes of the species that occur on that particular source. It is also a free source. Even if only a small quantity is gathered, the seed can be mixed with commercial supplies to improve the ecological value of the restored area.

Most seeds should be gathered during July, August and September, but timing will vary with species and locality. Grass seed can be collected by cutting the heads as they ripen, and leaving them spread on trays or paper in a dry place. The seeds can be shaken off when dry. Other plants may be more difficult, as the seeds on each head may not ripen simultaneously. The best method is to hold a plastic tray or container under the plant and gently shake the stem, or rub the seed head between the finger and thumb. This ensures that only the ripe seed is collected, and there should then be no danger of depleting the natural stock. Always consult the Nature Conservancy Council or local naturalists' trust if there is any doubt over the wisdom of collecting any species which may be uncommon in that locality.

Leguminous seeds should be gathered just before maturity as the pods of some species explode and scatter the seeds. Gather the pods, and cover loosely to prevent the seeds being lost when the pods explode. Store seed in paper bags or cardboard boxes, but not in polythene bags. Seed can be sown any time between March and September, although April and May are the optimum months.

Local groups, or people responsible for sites over the long term may find it worth establishing stock plants of native species, in the same way that commercial seed houses operate. These stock plants are grown in a nursery in optimum conditions, and so can produce far higher yields than if competing in the wild. For example, one cowslip plant can produce 4,000 seeds (4.54g) in a season. For further details of seed collection see Wells (1981).

Litter and sweepings

Sweepings and plant litter of various sorts may contain suitable seed. Gorse and heather litter can be gathered and spread like a mulch on bare patches of heathland to encourage regeneration. Hay bales may contain suitable seed, ideally cut from the type of sward which you wish to produce. The seed can be extracted by either beating or trampling the hay over a sheet or clean floor, or the hay can be spread directly onto the site.

Sweepings from hay barns also contain seed, but should not be used on ecologically sensitive sites as they could contain undesirable agricultural strains. Seed merchants may give away sweepings but the species composition will be variable.

SOIL STABILIZATION

Sprays

Various products are available which are applied by knapsack sprayer or watering can onto slopes after seed and fertiliser have been spread. They bind the surface and prevent seed being washed or blown away. Products include bitumen and rubber emulsions, and seaweed extracts. Suppliers are listed on page 182 (11.5 - 11.7).

Netting

Wyretex netting (see p79) can be pinned to the substrate or soil, and then covered with soil and seed. The grass then roots through the fabric to produce a tough and trample-resistant sward. Problems have been experienced using this technique on steep slopes, as the soil tends to slide down the netting. In these situations the Wyretex should be pinned over the topsoil, and then the seed scattered over it, preferably with a stabilizing spray. Trampling should be excluded from the area for at least a year.

Chicken wire (25mm mesh) can be used in the same way, and tends not to be as 'slippery' on steep slopes, so that soil and seed can be spread over the top. The wire is thus buried beneath the surface of the turf, and there is less danger of it becoming exposed. However, the wire does rust and break down after a few years, and so does not provide a long term strengthening in the way that Wyretex does.

Any turf that is strengthened with netting and subject to trampling must be checked periodically and repaired as soon as any netting becomes exposed. Fence off the damaged section. Cover the exposed parts with soil and rake seed and fertiliser into it, and into any other patches that look suspect.

Broplene Land Mesh (p182 11.8) is a 25mm polypropylene mesh used mainly for strengthening existing turf, but it can also be used in the way described above.

Brashings

Cut scrub can be spread over newly sown areas and slopes to protect the soil and seed from being eroded, and to deter trampling. Do not spread so thickly that seedlings are shaded out.

REQUIREMENTS OF SEED MIX

The choice of seed mix will depend on the physical nature of the site, and the use to which it will be put after restoration. The mixture is determined by the percentage of different species and strains. A cultivar is a named strain of a species, maintained in cultivation, and can be very different from other cultivars of the same species. As explained below, it can be important to use a certain cultivar. Cultivars are distinguished by a name or letter and number after the species name. The following characteristics should be considered when selecting a mix for seeding.

Nurse species

The mix should include a species which grows rapidly to protect the soil and other seed from being washed away, and provides a suitable microclimate for the germination and growth of other species. This is usually perennial rye grass (Lolium perenne) which establishes quickly, especially if fertiliser is applied, but declines as fertility decreases. It should not however be used on sites of high ecological value, such as National Nature Reserves or Local Nature Reserves.

At altitudes over 500 metres it is recommended that dwarf timothy (Phleum bertolonii S50) be used instead of perennial rye grass.

Resistance to trampling

It is unlikely that re-seeded areas will be kept entirely free of trampling, whether by animals or humans. Some trampling will be desirable if the aim is to maintain a grass sward and prevent it being invaded by scrub. Species resistant to trampling may therefore be selected, and will include those with tough leaves, often in a rosette, a growing point well below the soil surface, and an ability to propagate by stems as well as by seed. Footpaths tend to develop this type of species composition, as non resistant species are destroyed by trampling and resistant ones take their place. These differences can be seen quite clearly where path vegetation shows as a different colour, or resistant species such daisy (Bellis perennis) are in flower along the line of a path.

Species resistant to trampling are listed in the table on page 141.

USEFUL PLANT CHARACTERISTICS.
A. Rosette.   B. Rooting stems.

Ribwort Plantain.   Creeping Bent.
(Plantago lanceolata) (Agrostis stolonifera)

c. Nitrogen Fixing.   D. Strong tufted growth.

Black Medick.
(Medicago lupulina)

Crested Dog's Tail (Cynosurus cristatus)

Level of productivity

The mix should be productive enough to give a cover that will prevent soil erosion, but not so thick that local species are unable to recolonise. For example, the standard Department of the

Environment road verge mixture contains Lolium perenne S23 and Festuca rubra S59 which create a dense sward that makes colonisation by other plants difficult. These cultivars should be avoided if you wish to restore a semi-natural plant community. Fertilising the ground can also slow the return of native species because their growth rates cannot usually compete with those of sown species while soil fertility is high.

Colour

'Improved' grassland, containing agricultural grass species, is a brighter green than unimproved semi-natural grassland. Restored areas may be visible for some years as a bright green 'scar' if certain strains are used. Bright green looks especially out-of-place amongst the sombre colours of moorland, or on mountain and down-land slopes.

GROUND PREPARATION

Chemical treatments

If a large area is to be treated which may involve substantial expenditure on seed, it is essential to have the soil analysed so that any requirement for lime or fertiliser can be calculated. The local office of the Ministry of Agriculture, Fisheries and Food can arrange for soil tests to be done, and will often give advice on suitable seed mixes. In Scotland, contact the Department of Agriculture and Fisheries, or one of the agricultural colleges.

Many eroded slopes have no soil left, and in the absence of available topsoil, seeding must be done direct onto a loose stony substrate. Analysis of these substrates is likely to be meaningless as nutrient levels are so low, although the pH can be checked to find out if liming is necessary. The example below (p136) and the previously mentioned example of Nab Scar, Cumbria, show that successful seeding is possible even on very stony substrates with no soil.

Tilth

Preparing a seed bed is very important, and is worthwhile even where the site looks beyond redemption. Soils on flat areas will be very compacted, and should be broken up with a fork or pick. Ideally this should then be left to be broken further by weathering, but normally this will have to be done straightaway, by knocking the lumps apart with the fork or rake. Remove stones, and rake to produce as fine a tilth as possible. Any remaining soil on slopes should be raked to break the surface crust, but do not loosen more than this, or you may encourage further erosion.

Topsoil brought from elsewhere should be roughly spread with a rake, and then trodden to firm it. Tread in an orderly plan as shown below. Rake the surface again to produce the final tilth.

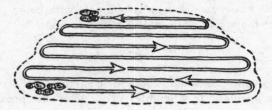

Loose stony substrates without soil can still be raked to remove large stones, and to produce as even a surface as possible. Always rake slopes from the bottom upwards. If any mulch such as sewage sludge or composted bark is being used on stony slopes, it must be well raked in so that the mulch and substrate are mixed to at least 100mm. Otherwise the plants will merely root into the surface mulch, which will slide away like a carpet after heavy rain or snow.

SITE STUDIES

The following examples give guidelines for seed mixes and treatments for various situations. It will be noted that seeding rates vary between examples. For most purposes, a rate of between 5 and 15g/sq m (50-150kg/ha) will be sufficient.

Coastal cliffs

	%wgt
Red fescue (Festuca rubra)	38
Yorkshire fog (Holcus lanatus)	35
Creeping bent (Agrostis stolonifera)	17
Ribwort plantain (Plantago lanceolata)	5
Great plantain (Plantago major)	4
White clover (Trifolium repens)	0.6
Sea plantain (Plantago maritima)	0.4

This mix was successfully used at Kynance Cove, on the Lizard Peninsular, to restore an eroded area of cliff top. Local topsoil was brought in, and trampling was excluded by the construction of a new path. Locally collected seed was used in addition to the commercial seed listed above, and germinated slightly better than the commercial seed. The percentages quoted above are merely a record of what was used in this case, and are not necessarily significant. Different proportions plus additional suitable species could be used according to availability.

Saline areas lower down the cliff did not establish successfully. This very extreme type of location would require seeds collected from the immediate area to ensure they were suitable maritime ecotypes, and the complete exclusion of trampling. For further details of the scheme see 'Kynance Cove' (O'Connor, Goldsmith and Macrae, 1979)

Calcareous soils

The following is the Countryside Commission recommended 'Downland Mix'.

	%wgt
Perennial rye grass (Lolium perenne)	30
Red fescue (Festuca rubra)	20
Chewing's fescue (Festuca rubra commutata)	20
Dwarf timothy (Phleum bertolonii)	15
Creeping bent (Agrostis stolonifera)	10
White clover (Trifolium repens)	5

Calcareous subsoil

The seed mix given below has been used for the reinstatement of vegetation on raw calcareous subsoil, thus obviating the need for topsoil (Wathern and Gilbert, 1979). The absence of perennial rye grass from the mix slows the initial establishment of the sward, but allows the quicker invasion of local species. The mix should be sufficient to prevent erosion, and has a pleasant grey-green appearance. Fertiliser is not necessary. The recommended seeding rate is 12.5g/sq m.

	%wgt
Sheep's fescue (Festuca ovina)	40
Red fescue (Festuca rubra rubra)	20
Chewing's fescue (Festuca rubra commutata)	10
Smooth meadow grass (Poa pratensis)	10
Common bent (Agrostis tenuis)	10

Stony brown earth

	%wgt
Red fescue (Festuca rubra)	35
Timothy (Phleum pratense)	35
Crested dog's tail (Cynosurus cristatus)	15
Smooth meadow grass (Poa pratensis)	15

This mix was used on sloping ground at Tarn Hows, Cumbria, which is 180 metres above sea level. Various trials were carried out using different combinations of topsoil, seed and fertiliser. Topsoil, where used, was laid 150mm thick with lime at 250g/sq m. Fertiliser (NPK 12:18:18) was applied to some plots at 50g/sq m. After three years it was found that, in the absence of trampling, all plots recovered equally, including those that received no treatment. In this situation there was plenty of seed spread from surrounding vegetation, but seeding was considered worthwhile for cosmetic reasons, and to prevent soil erosion. For further details of the scheme see 'Tarn Hows' (Brotherton, Maurice, Barrow and Fishwick, 1977).

Very disturbed ground

	%wgt
Perennial rye grass (Lolium perenne)	55
Red fescue (Festuca rubra)	25
Common bent (Agrostis tenuis)	5
Smooth meadow grass (Poa pratensis)	5
Crested dog's tail (Cynosurus cristatus)	5
White clover (Trifolium repens)	5

This seed mix has been successfully used on derelict urban sites with little or no topsoil, by using heavy applications of fertiliser. The technique was developed for short-term treatment of sites awaiting development, but could also be used on eroded, stony ground, where there is no possibility of obtaining topsoil. Liming will probably be necessary, especially on upland sites. Seed should be sown at 8g/sq m, followed by an application of fertiliser (NPK 15:15:15) at 60g/sq m, both raked in. Extra applications of fertiliser may needed in the following year or two, but in the absence of heavy trampling, the sward should slowly improve (Bradshaw and Handley).

This technique should not be used where fertiliser run-off could damage important plant communities, including aquatic vegetation.

Acidic peat

The natural plant communities of acidic peat are rapidly destroyed by trampling. The establishment of a grass sward is difficult, and should only be attempted where it is considered a preferable alternative to hard surfacing as a means of providing a trample-resistant surface. For successful establishment, the peat must be analysed to determine lime and fertiliser requirements for each site.

The establishment of grass is only recommended for sites where the plant communities are already degraded, as there is a danger of lime and fertiliser leaching out of the treated area and damaging the surrounding flora. Lime and fertiliser application must be done with great care, to ensure that the surrounds do not get contaminated. As well as being ecologically out-of-place, grass, by its colour, can look obtrusive in the heathland or moorland landscape.

This seed mix was used at Risley Moss, Warrington, to establish grass in clearings in acidic woodland. The pH was raised to 5.5 by liming at 0.8kg/sq m. This should be done immediately before seeding as lime is rapidly leached from the soil. Phosphate levels were also low, so 'Basic Gafsa' (P_2O_5 content 29%) was used at 10g/sq m. This is not lost by leaching and is best applied several weeks before sowing. A general

fertiliser (NPK 15:15:15) was also used at 38g/sq m. Trampling was excluded for at least a year after seeding (J D Moffat; unpublished paper).

	%wgt
Chewing's fescue (Festuca rubra commutata)	30
Red fescue (Festuca rubra rubra 'Boreal')	20
Sheep's fescue (Festuca ovina 'Novina')	20
Common bent (Agrostis tenuis 'Highland')	15
Annual meadow grass (Poa annua)	10
White clover (Trifolium repens 'Huia')	5

The mix below was used on heavily trampled peat at Beacon Fell, Lancashire. Tests showed that trampling further lowered the pH, as well as causing compaction and the formation of anaerobic conditions, so giving no chance of natural recovery even if trampling was excluded.

	%wgt
Annual meadow grass (Poa annua)	60
Red fescue (Festuca rubra rubra)	25
Common bent (Agrostis tenuis)	15

After drainage of waterlogged areas, limestone was applied at 2kg/sq m, and rotovated into the peat. Fertiliser (NPK 17:17:17) was then applied at 75g/sq m, followed by grass seed at 20g/sq m. The seed bed was then rolled.

Trampling should be excluded for at least four months, and the grass cut and rolled as necessary to encourage the development of a resistant sward (Gemmell and Crombie, 1978).

Mountains

Extremes of climate, steep slopes and thin soils reduce the number of species able to survive on mountains, and slow the rates of growth and seed production. Natural recovery of damaged ground occurs only very slowly.

Establishment of trample-resistant turf on paths is unlikely to be possible, and efforts at revegetation should be directed to damaged areas where trampling can be reduced or excluded. The aim of seeding is to prevent further erosion and provide a microclimate which encourages invasion by local species. Up to about 500m, a wide range of annual and perennial species should invade. Heather (Calluna vulgaris) has been found to invade sites up to 800m in the Cairngorms, in the absence of trampling. Above this height, recolonisation is slower, and is dominated by mosses. Seeding has been successfully done to an altitude of 1100m in the Cairngorms (Bayfield, 1980). The following seed mix can be used on sites up to 800m, at a minimum rate of 9g/sqm. This should be increased to 15g/sq m on steep slopes or sites without topsoil.

	%wgt
Red fescue (Festuca rubra)	45
Crested dog's tail (Cynosurus cristatus)	20
Smooth meadow grass (Poa pratensis)	20
Common bent (Agrostis tenuis)	15

Small amounts of the following species can be added to the mix, for use up to altitudes of about 500m.

Yarrow (Achillea millefolium)
White clover (Trifolium repens)
Bird's foot trefoil (Lotus corniculatus)
Ribwort plantain (Plantago lanceolata)

The following seed mix is recommended by the Countryside Commission for sites above 350m.

	%wgt
Red fescue (Festuca rubra rubra)	45
Dwarf timothy (Phleum bertolonii S50)	25
Common bent (Agrostis tenuis)	15
Smooth meadow grass (Poa pratensis)	15

Prior to seeding, rake in limestone at 250g/sq m, fertiliser (NPK 20:20:20) at 38g/sq m, and phosphate at 100g/sq m. Repeat the fertiliser application the following year if growth is poor. Seed should be sown with a stabiliser (see p134) to prevent it being blown away.

Turfing

Turf has been successfully used in many situations to repair damaged ground, from coastal cliffs to mountain slopes at 1200m. Moving turf is possible at any time of year, but should not be attempted during a prolonged drought.

Where a path has braided, leaving isolated patches and strips of turf, these can be removed and re-used along the sides of the path.

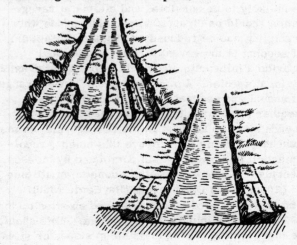

Extra turf will usually be needed, and should be dug from the nearest place where its removal will not be too obtrusive in the landscape, and where there is no possibility of erosion occurring before the cover has re-established. The turf must be of a type that will blend with the surrounding vegetation, and fulfill any function required, such as being resistant to trampling.

Suggested sites are as follows:

a   It may be possible to take turves from the edge of an adjoining field, close to the wall or hedge. Permission must of course be gained from the landowner.

b   On open hillsides of rough vegetation, turves can be taken at random. Try to take the turves away from the lines along which walkers may go, as inevitably quite deep pits will be left which could cause a walker to stumble. Turves of heather and rough grass are difficult to cut thinly, and you will probably have to remove a sod about 150mm thick, which leaves a potentially dangerous hole.

c   In order to avoid creating these scattered holes, it is often better to cut a ditch. As well as being less hazardous to walkers, this also minimises the disturbance to the ground during turf-cutting. This is very important on wet moorland, which is easily damaged by trampling.

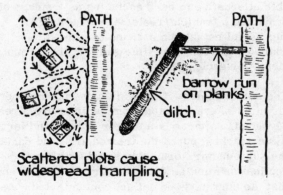

Scattered plots cause widespread trampling.

Often the ditch can be part of a drainage scheme for the path, but if not, align the ditch along the contours so that is does not act as a drain. The ditch will gradually fill as the sides collapse and blown material collects.

d   If there is an expendable piece of ground nearby it may be permissible to strip the turf in commercial fashion. If possible, fill and reseed it to hasten recovery. Cut the turves as neatly as possible to avoid wastage. The standard horticultural turf is 300mm x 900mm, which gives a guide to the size that is reasonable to handle. A crawler tractor will cut turf in long strips which should be cut into manageable pieces, or it will only be broken in transit.

Protecting the turf

When replacing turves on slopes, it may be necessary to protect them in some way to prevent them slipping before they have become established. The same device may also serve to keep people or animals off.

a   Pin 25mm chicken wire over the turf, using loops of galvanised fencing wire. Pin at least every 200mm along the edge to discourage dogs and other animals burrowing underneath. Public access should be prevented, or the chicken wire will rapidly be exposed and the erosion problem recur. The chicken wire will eventually rust away.

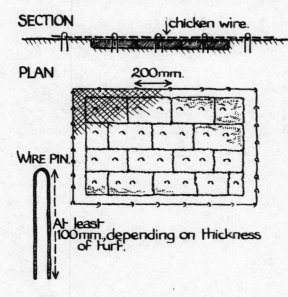

SECTION | chicken wire.

PLAN   200mm.

WIRE PIN.

At least 100mm, depending on thickness of turf.

b   At Badbury Rings, Dorset, chicken wire was pinned using ash pegs cut locally. This discourages access, but does look rather unattractive.

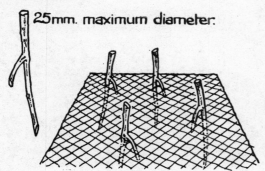

25mm. maximum diameter.

c   Wyretex (see p79) or Broplene Land Mesh (see p134) can be pinned in the same way as chicken wire, but being polypropylene, will last longer. These products are suitable for areas which will be heavily trampled.

The above three methods are only suitable for close-cut turves, as the mesh must be in close contact all over the surface of the turf, for it to bind securely.

d   Brashings can be spread over the turves to discourage trampling, and staked to prevent slippage. This method is best used on rough turves. The brashings can be left in place to rot down, or removed in winter when grass growth is low, and there is less chance of tearing up any turf with the brashings.

Transplanting

Transplanting tussocky vegetation, shrubs and small trees can be useful for the following reasons:

a   They can act as a barrier and discourage trampling.

b   The canopy intercepts rainfall, and so reduces direct run-off on slopes.

c   The roots help bind the soil and prevent erosion.

d   Transpiration draws up water through the root system, so lessening sub-surface flow which contributes to slope instability.

e   The canopy softens and disguises the appearance of damaged slopes.

Bracken

This grows from a dense mat of rhizomes, and can be used to stabilize ground and discourage trampling. It is not easy to dig up as the rhizomes grow very deep in the soil, but can be moved in winter, when the plant is dormant. Only use where successful establishment will not itself become a problem, as bracken is difficult to eliminate.

Gorse

This is a useful plant for discouraging access, and young plants can be moved successfully, preferably during the winter or early spring.

Hawthorn and blackthorn

Young plants can be transplanted and should
survive even in poor stony soils and in exposed
situations.

Bog myrtle

This is a shrubby plant of upland areas, which
grows in wet ground. Attempts to transplant it
at Cwm Idwal, North Wales, have not so far been
successful, but as the plant is not eaten by sheep,
further experimentation would seem to be worthwhile.

Heather

Young heather plants can easily be transplanted,
preferably in early spring. At the Devil's Jumps,
Frensham, Surrey, heather seed and cuttings
were collected locally, grown on in a nursery, and
the young plants used to help stabilise a repaired
gully.

Sedges and rushes

Clumps of sedges and rushes make useful barriers,
as they are uncomfortable to walk over, and
further discourage by being associated in people's
minds with wet ground. This technique was used
at Cwm Idwal and succeeded in keeping people to
the path where fences and barriers had failed.

Nursery grown stock

In some situations it may be worth buying in
plants to use for barriers and slope stabilization.
Species chosen should be native to the area. Useful characteristics include prickly growth to
deter browsing and trampling, a deep tap root for
anchorage on unstable slopes, and the ability to
withstand water and nutrient stress. Some
species are listed in the table on page 145.

Planting should only be considered if proper care
can be given in the first few years. Most sites
will have to be fenced against sheep and rabbits.
Fencing does have the advantage of keeping
walkers off the area for a reason that they can
see is legitimate, and will hopefully respect.
Explanatory notices should be used. Some
weeding and trimming will be necessary inside
an ungrazed enclosure, as well as periodic
checking of fences, stakes and tree ties.

TABLE 11a GRASSES AND HERBACEOUS PLANTS RESISTANT TO TRAMPLING

* Denotes grass
The range of heights given is for plants not heavily trampled or grazed

Name	Height (mm) and habit of growth	Habitat and distribution	Reproduction	Comments
Achillea millefolium (Yarrow)	100 – 300	Widespread and common in grassland	Stolons and seed	
Agrostis canina * (Velvet bent)	150 – 750 Tufted	Widespread in damp places, tolerates shade	Stolons and seed	
Agrostis stolonifera * (Creeping bent)	80 – 400 Tufted	Widespread up to 750m	Stolons and seed	
Agrostis tenuis * (Common bent)	100 – 700 Tufted	Abundant and widespread to 1230m	Rhizomes and seed	Especially dominant on dry acid soils
Bellis perennis (Daisy)	20 Rosette	Abundant in short grassland to 900m	Seed	
Bromus erectus * (Upright brome)	40 – 1200 Densely tufted	Well-drained calcareous soils in southern England	Seed	
Bromus inermis * (Hungarian brome)	50 – 1000 Tufted	Introduced. Naturalised on sandy and stony soils.	Rhizomes and seed	Extensive rhizome system. Drought resistant. Stabilizes slopes.
Carex flacca (Glaucous sedge)	100 – 400	Widespread and common on calcareous soils.	Seed	
Convolvulus arvensis (Bindweed)	Climbing and scrambling	Widespread and common	Seed	Very deep roots
Coronopus squamatus (Wart cress)	Prostrate	Wasteground. Rare in north	Seed	Annual/Biennial
Crepis capillaris	Rosette	Common and widespread on grass, heaths, wasteground.	Seed	Annual
Cynosurus cristatus * (Crested dog's tail)	50 – 750 Compactly tufted	Common and widespread to 600m. Poor competitor, but tolerant of low fertility.	Seed	Withstands drought and cold. Stays green in winter. Good sheep grazing.

Name	Height (mm) and habit of growth	Habitat and distribution	Reproduction	Comments
Dactylis glomerata * (Cocksfoot)	150 – 1400 Densely tufted	Abundant and widespread in pastures and rough grassland	Seed	Nutrient demanding. Avoid sowing with young trees or shrubs.
Festuca ovina * (Sheep's fescue)	50 – 600 Densely tufted	Widespread on poor acidic and basic soils, up to 1230m.	Seed	Slow to germinate. Hardy and drought resistant. Withstands heavy grazing.
Festuca rubra * (Red fescue)	150 – 900	Widespread and abundant to 1230m, esp on acid soils.	Seed	Slow to establish. Browns in winter. Intolerant of close cutting. Numerous variants.
Festuca rubra commutata * (Chewing's fescue)	150 – 900	Occurs naturally on well drained soils in south. Common in lawns.	Seed	Fine leaved and drought resistant.
Galium saxatile (Heath bedstraw)	100 – 200 Mat-forming	Common and widespread on acid soils.	Seed	
Hieracium pilosella (Mouse-ear hawkweed)	100 Rosette	Grassy pastures and heaths	Rhizomes, stolons and seed.	
Holcus lanatus * (Yorkshire fog)	200 – 1000 Tufted	Very common in wide range of soils and situations.	Seed	
Holcus mollis * (Creeping soft grass)	200 – 1000 Tufted	Widespread. Tolerant of shade.	Rhizomes and seed	Very extensive rhizomes. Troublesome weed of sandy fields.
Hypochaeris radicata (Cat's ear)	70 – 250 Rosette	Common and widespread in grassy dunes and waysides.	Seed	
Juncus squarrosus (Heath rush)	80 – 150	Acid moors, bogs and wet heaths	Seed	
Koeleria cristata * (Crested hair-grass)	100 – 600 Compactly tufted	Dry calcareous grasslands and shady places	Rhizomes and seed	
Leontodon hispidus (Rough hawkbit)	Rosette	Meadows and pastures, esp on calcareous soils	Seed	
Leontodon taraxacoides (Hairy hawkbit)	Rosette	Dry grassland on basic soils. Sand dunes.	Seed	
Lolium perenne * (Perennial rye grass)	100 – 900 Tufted	Old pastures and meadows, esp on heavy soils. Declines as fertility decreases.	Seed	Extensively sown for pastures. Rapid establishment and long growing season. Many different strains.

Name	Height (mm) and habit of growth	Habitat and distribution	Reproduction	Comments
Lotus corniculatus (Bird's foot trefoil)	100 – 400 Spreading	Widespread in grassy places	Seed	
Luzula campestris (Field wood rush)	10 – 150 Loosely tufted	Very common in pastures	Stolons and seed	
Medicago lupulina (Black medick)	50 – 500	Common in grassy places, but not on very poor soils.	Seed	Nitrogen fixing
Nardus stricta * (Mat grass)	100 – 400 Densely tufted	Widespread in most soils and on mountains to 1000m.	Rhizomes and seed	Tough and wiry. Not palatable to sheep or cattle.
Phleum bertolonii * (Dwarf timothy)	100 – 500 Tufted	Old pastures and downlands	Stolons and seed	Many different strains.
Plantago coronopus (Buck's-thorn plantain)	Rosette	Light soils and rock crevices near the sea.	Seed	
Plantago lanceolata (Ribwort plantain)	Rosette	Widespread and common except on very poor soils.	Seed	
Plantago major (Great plantain)	Rosette	Widespread in cultivated ground	Seed	
Poa annua * (Annual meadow grass)	30 – 300 Tufted	Widespread in most soil types to 1000m.	Seeds throughout year.	Invades bare ground and sown grass. Bright green. Pollution resistant.
Poa subcaerulea * (Spreading meadow grass)	100 – 400	Marshy pastures, moist hill slope to 600m, esp in north.	Rhizomes and seed	Extensive rhizome system
Poa pratensis * (Smooth meadow grass)	100 – 900 Tufted	Widespread and common in soils to 1230m	Rhizomes and seed	
Poa trivialis * (Rough meadow grass)	200 – 1000 Loosely tufted	Widespread on rich moist soils. Tolerant of partial shade.	Stolons and seed	Purplish-green
Polygonum aviculare (Knotgrass)	Prostrate	Wasteground, arable field and sea shores.	Seed	
Potentilla anserina (Silverweed)	Rosette	Common in damp pastures, dunes and wasteground, to 500m.	Stolons and seed	
Potentilla reptans (Creeping cinquefoil)	Rosette	Wasteground and waysides on basic and neutral soils, to 500m.	Stolons and seed	

Name	Height (mm) and habit of growth	Habitat and distribution	Reproduction	Comments
Poterium sanguisorba (Salad burnet)	150 – 400	Widespread and common in Eng. and Wales, on calcareous grass land to 500m.	Seed	
Prunella vulgaris (Self-heal)	50 – 300	Common in grassland, woodland clearings and wasteground.	Rhizomes and seed	
Rumex acetosella agg. (Sheep's sorrel)	20 – 300	Widespread and common except on poorest soils.	Rhizomes and seed	
Sagina procumbens (Procumbent pearlwort)	Rosette	Common on paths, lawns and banks.	Stolons and seed	
Sedum anglicum (English stonecrop)	20 – 50 Mat	Dunes, shingle, rocks, dry grassland to 1100m. Not on strongly basic soils.	Stolons and seed	Tinged red
Silene vulgaris (Bladder campion)	250 – 900	Widespread on grassy slopes, roadsides, arable land	Seed	
Taraxacum officinale (Common dandelion)	Rosette	Common in pastures, waste-ground and waysides.	Seed	
Thymus drucei (Thyme)	10 – 70 Mat	Common and widespread in dry grassland, heaths and scree.	Seed	
Trifolium pratense (Red clover)	600	Widespread and common in grassland	Seed	Nitrogen fixing
Trifolium repens (White clover)	500	Widespread and common in grassland	Stolons and seed	Nitrogen fixing
Trisetum flavescens * (Golden oat-grass)	200 – 800 Loosely tufted	Common in England on old pastures, especially on calcareous soils.	Seed	Drought resistant
Veronica chamaedrys (Germander speedwell)	200 Prostrate	Very common in grassland, woodland and hedgebanks.	Stolons and seed	

The above table is based on the list of plant species resistant to trampling in Speight (1973), with information from Hubbard (1954) and Clapham, Tutin and Warburg (1959).

TABLE 11b PLANTS FOR SLOPE STABILIZATION, SCREENS AND BARRIERS

| Type | Herbaceous perennial | Non-woody plant that dies down in winter. |
| | Perennial herb | Non-woody plant that retains its leaves in winter. |

Use	Slopes	Suitable for slope stabilization and protection.
	Screen	Planting that discourages access by blocking sight lines.
	Moorland	Restoration of eroded moorland.
	Barrier	Planting that physically prevents access.

| Comments | N fixing | Fixes nitrogen from the air, therefore useful on poor soil. |

Name	Type	Habitat	Use	Comments
Acer campestre (Field maple)	Deciduous tree	Exposed hillsides and banks on calcareous soils and clay.	Slopes Screen	Seeds and colonises quickly
Ajuga reptans (Bugle)	Herbaceous perennial	Damp woods, pastures, banks.	Slopes in shade	
Alnus glutinosa (Alder)	Deciduous tree	Damp soils, both acid and calcareous. Polluted soils.	Slopes Screen	N fixing. Good pioneer. Fast growth when young.
Anthemis nobilis (Chamomile)	Perennial herb Leaves persist	Sandy slopes and roadsides in England and Wales.	Slopes	
Arctostaphylos uva-ursi (Bearberry)	Evergreen shrub	Moors, rocky ground to 1000m.	Slopes Moorland	
Arrhenatherum elatius (False oat-grass)	Perennial grass	Rough grassland, roadsides, shingle and gravel banks.	Slopes	Rapid leafy growth. Drought resistant.
Calluna vulgaris (Heather)	Evergreen shrub	Heaths, moors, bogs and open woods on acid soils.	Slopes Moorland	Does not tolerate trampling.
Carpinus betulus (Hornbeam)	Deciduous tree	Chalk or clay soils. Withstands wind and shade.	Slopes	Not usually damaged by grazing animals.
Clematis vitalba (Travellers joy)	Deciduous climber	Hedgerows and woodlands on calcareous soil.	Slopes	
Cotoneaster microphyllus (Cotoneaster)	Evergreen shrub	Introduced. Naturalised, especially on limestone cliffs near sea.	Slopes Barrier	Very invasive
Crataegus monogyna (Hawthorn)	Deciduous tree	Tolerates wide range of soils except very wet or acid. Tolerates exposure.	Slopes. Screen and barrier.	

Name	Type	Habitat	Use	Comments
Cytisus scoparius (Broom)	Deciduous shrub	Heaths, open woods on acid soils.	Slopes. Screen and barrier.	N fixing.
Erica cinerea (Bell heather)	Evergreen shrub	Dry heaths and moors to 700m.	Slopes Moorland	Does not tolerate trampling.
Euonymus europaeus (Spindle)	Deciduous shrub/ small tree	Woods, scrub, waysides on calcareous soils.	Slopes. Screen and barrier.	
Frangula alnus (Alder buckthorn)	Deciduous shrub/ small tree	Damp, slightly acid or peaty soils in England and Wales.	Moorland. Screen and barrier.	
Fraxinus excelsior (Ash)	Deciduous tree	Rich damp soils, and calcareous soils. Withstands wind and exposure.	Slopes	Casts only light shade, so allowing growth of ground cover. Fast growing. Gross feeder.
Hedera helix (Ivy)	Evergreen climber	Shady places on all but very acid or waterlogged soils.	Slopes in shade.	
Hippophae rhamnoides (Sea buckthorn)	Deciduous shrub	Fixed dunes and sea cliffs.	Slopes. Screen	Thorny. N fixing.
Ilex aquifolium (Holly)	Evergreen shrub/ tree	Wide range of soils except very wet or very dry. Tolerates wind and shade.	Slopes. Screen and barrier.	Prickly.
Juniperus communis (Juniper)	Evergreen shrub/ tree	Chalk, limestone and slate soils. Tolerates wind.	Slopes. Screen and barrier.	Slightly prickly.
Lamium album (White deadnettle)	Perennial herb Leaves persist	Hedgebanks and waysides in England.	Slopes	
Larix decidua (Larch)	Deciduous tree	Tolerates very dry or waterlogged soils on cold exposed sites.	Slopes Screen	Good pioneer. Rapid growth in early stages.
Lathyrus latifolius (Everlasting pea)	Herbaceous perennial	Introduced. Naturalised on banks and hedgerows.	Slopes	N fixing
Ligustrum vulgare (Privet)	Evergreen shrub	Hedges and scrub on calcareous soils. Widely naturalised elsewhere.	Screen and barrier.	
Lonicera periclymenum (Honeysuckle)	Climber	Woods, hedges, scrub, shady rocks to 600m.	Slopes	
Origanum vulgare (Marjoram)	Perennial herb Leaves persist	Dry pastures and hedges. Usually calcareous soils.	Slopes	

Name	Type	Habitat	Use	Comments
Pinus sylvestris (Scots pine)	Evergreen tree	Tolerant of exposed mountain and coastal sites.	Slopes	Deep tap root. Intolerant of pollution. Invades on heaths.
Populus alba (White poplar)	Deciduous tree	Introduced. Naturalised in south. Tolerant of exposed sites, wet soils.	Slopes Screen	Suckers freely.
Populus tremula (Aspen)	Deciduous tree	Woods, espcially on poor soils in north and west.	Slopes Screen	Suckers freely
Prunus spinosa (Blackthorn)	Deciduous shrub/ tree	Scrub and hedges on most soils to 450m.	Slopes. Screen and barrier.	Thorny. Suckers. Can be very invasive.
Quercus petraea (Sessile oak)	Deciduous tree	Woods throughout Britain, and esp on acid soils in north and west.	Slopes	
Quercus robur (English oak)	Deciduous tree	Dominant tree in lowlands on basic loams and clays; to 450m.	Slopes	Faster growth than commonly supposed.
Rosa canina (Dog rose)	Shrub	Woods, hedges, scrub, on most soils.	Slopes Barrier	Prickly. Many species and variants.
Salix alba (White willow)	Deciduous tree	Rich damp soils.	Slopes. Screen and barrier.	Rapid growth. Forms thick barrier if coppiced.
Salix caprea (Great sallow)	Deciduous shrub/ tree	Woods, scrub and hedges to 900m.	Slopes. Screen and barrier.	
Sambucus nigra (Elder)	Deciduous shrub	Scrub and waysides, esp on base-rich, N-rich and disturbed soils.	Slopes	Rabbit resistant.
Sorbus aria (Whitebeam)	Deciduous tree	Chalk and limestone in southern England. Tolerates wind.	Slopes	
Sorbus aucuparia (Rowan)	Deciduous tree	Woods, scrub, rocks. Light soils. Northern Britain.	Slopes	Casts only light shade.
Ulex europaeus (Gorse)	Evergreen shrub	Heaths and light soils.	Slopes. Screen and barrier.	Prickly. N fixing.
Viburnum lantana (Wayfaring tree)	Deciduous shrub/ tree	Hedgerows on calcareous soils. Common in southern England.	Slopes. Screen and barrier.	Can be invasive.
Viburnum opulus (Guelder rose)	Deciduous shrub	Woods and scrub on damp soils.	Screen.	
Vinca minor (Lesser periwinkle)	Perennial herb Leaves persist	Woods and hedgebanks on most soils.	Slopes in shade.	

12 Stiles and Gates

Stiles, gates and barriers on paths serve two purposes:

1   To prevent stock straying.

2   To allow access for permitted path users, whilst excluding others.

As explained in Chapter 2, stiles and gates on rights of way are the responsibility of the land-owner, although a minimum grant of 25% is available from the local authority, in England and Wales. Stiles and gates in National Parks, Heritage Coasts and other areas being managed for recreation are usually provided and installed by the authority concerned. In other areas, voluntary groups have become involved in the provision of stiles, as the current grant system is not sufficient to encourage many landowners to maintain their stiles.

This chapter gives recommended dimensions for stiles, gates and barriers, with suggested designs, materials and methods of construction. It does not advocate standardisation, as much of the character of paths is given by the range of regional, local and individual designs. However, repeated use of a design increases efficiency for any group undertaking a lot of stile replacement. Complete fabrication to a standard pattern is not always advisable, because each site differs according to the slope of the ground, the ease with which posts can be put in, the condition of the fence, wall or hedge, and materials to hand.

Choosing a Stile

Durability

It is preferable to design for maximum durability, both for safety, and to make best use of the time and effort which goes into liaison with landowners, transporting and erecting stiles. For this reason, treated timber and a jointed construction are recommended.

Type of use

The majority of paths are used by fit and able walkers, and a basic stile design is quite adequate. However, special provision may need to be made on paths frequently used by the elderly, disabled, or mothers with pushchairs. Favourite walks in and around towns and villages, and popular coastal or hillside walks should not have stiles put in which exclude any of these people. Extra steps and handholds on stiles may be sufficient, or stiles can be replaced by kissing or bridle gates.

Conflict can easily arise, as bridle gates negotiable by wheelchairs can also give access for motor-bikes. Gates which exclude motorbikes are described on pages 164-166. Kissing gates are not easily negotiated by walkers with large ruck-sacks. Ideally, popular routes such as the valley paths in the Lake District need adjoining kissing gates and stiles, to allow a short stroll for the less fit, and access to the mountains for the backpackers.

Overweight or elderly dogs also find problems with stiles, particularly if their owners are likewise. 'Dog latches' have been provided on some popular dog-walks, to discourage dog owners from damaging fencing in order to get the dog through. Although these points may seem trivial to the agile walker, they make all the difference to the availability of a pleasant daily walk for a less able person.

Frequency of use

On very popular paths and those used by large groups of people, it may be worth putting in two stiles or kissing gates side by side to lessen 'traffic jams' and prevent the otherwise inevitable damage to walls or fences by those too impatient to wait. This has been done, for example, on the path to Malham Cove in Yorkshire.

Use also affects the durability of a stile. A frequently used one must be a top-quality job to withstand wear, but on less popular paths a cheaper job may be sufficient. A new, even if rough, stile seen from a roadside or path is the best inducement to get a newly cleared or way-marked path into use.

Transport and access for materials

As many stiles are put in by voluntary groups they must be of a size and design which can be easily transported on a roof-rack or in an estate car, and can be carried some distance by hand if necessary. Consider this before deciding on pre-fabrication.

Stock in field

A four bar stile may be necessary for a sheep-proof fence, or sheep netting can be attached to the stile. Some horses and ponies may learn to jump the standard 900mm height stile, and a higher one may have to be installed.

Access for machinery

Some stiles have been designed with removeable bars so that machinery can be taken through for annual clearance. Some of the stiles on the Offa's Dyke path were constructed in this way, but have not proved a popular design with walkers as they are awkward to climb. With the advent of portable brush cutters, this facility should become unnecessary.

<u>Existing fence or hedge</u>

Although this is totally the responsibility of the landowner, the stile must be positioned and constructed in such a way that the fence on either side is secure and stock-proof. Stile posts should not be used as fence-straining posts. An advantage of the ladder stile is that it is independent of the fence or wall it crosses, and the stile erector can safely ignore the condition of the boundary.

Stile Ergonomics

The basic dimensions, as recommended by British Standard 5709:1979 amended 1982, are as follows. These dimensions apply only to 'step-over' stiles. 'Step-through' and ladder stiles are different in principle (see page 156).

Width between uprights	1000mm min
Height of top rail above ground	900mm min
Height of bottom step above ground	300mm max
Rise between steps	300mm max

The maximum rise between a step and the top rail is therefore 600mm. The 300mm maximum height from the ground or between steps is an easy step for most people, aided if necessary by grasping the handhold. The width is important to allow the walker space to swing the leg sideways over the stile, without having to lift the knee higher than a maximum of 600mm. A gap narrower than this means the knee has to be lifted higher, which is a difficult action for the less agile.

The handhold on the stile post helps the walker to pull himself over the top rail, and to steady himself while climbing over. It should be about 1000mm above the level of the top step. A separate step is strongly recommended, as it allows the walker to face forward throughout the action of climbing over. A step slightly angled as shown is comfortable for most people, as it allows the right leg to lead over the top rail, an action which is natural to most right-handed people. The rhythm of walking is then not broken.

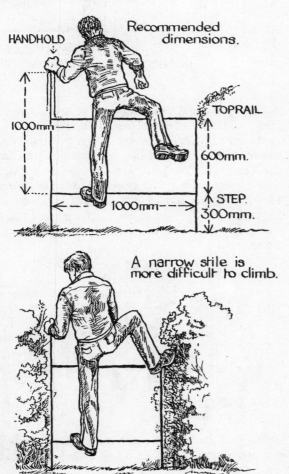

HANDHOLD   Recommended dimensions.

1000mm   TOPRAIL

600mm.

1000mm   STEP. 300mm.

A narrow stile is more difficult to climb.

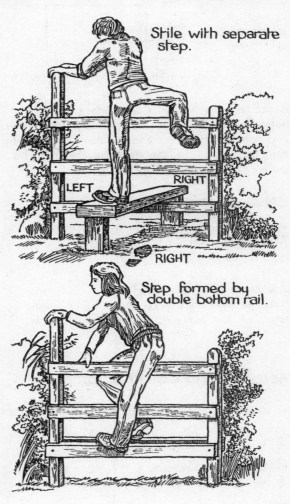

Stile with separate step.

LEFT   RIGHT

RIGHT

Step formed by double bottom rail.

A step formed by the bottom rail is not easy to negotiate, as one has to turn while crossing the stile, and there is often not enough room on the rail for a secure foothold. The rail can be dangerous as a foothold if wet or muddy.

Two steps will be needed if either:

a   The stile is on sloping ground.

b   The top rail needs to be higher than 900mm to keep stock in.

The top rail should not be more than 450mm above the upper step.

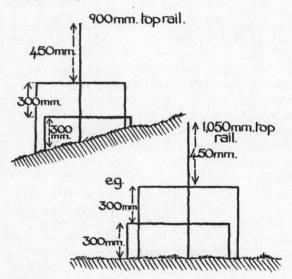

It can be awkward to fit two steps comfortably between the gaps in the rails. Only pre-drill the rail positions if careful measurement has been done of the site, as the rails may prove to be in the way of one or other of the steps.

Structural dimensions

The basic structural dimensions for step-over stiles are as follows. Timber sizes are given for each design.

Maximum space between rails	300mm
Depth of stile post in ground	750mm
Depth of step support in ground	500mm

Materials and Construction

See page 88 for information on joints and fixings.

Wood

Tanalised or celcurised softwood is recommended (European larch, pine, Douglas fir, Western red cedar). See page 183. Softwoods are lighter to transport than hardwoods, and are easier to cut, drill and nail. Hardwoods are more durable, but are better used only if tools and transport are available to easily deal with it.

Joints

BS 5709:1979 recommends mortised joints for rails. These are strong, but need special tools to construct. The rails are also difficult to replace if vandalised or accidentally broken, without loosening the fixing of the stile posts. Mortises can be made by the timber merchant, and are best assembled in the workshop, drilled and dowelled, and taken prefabricated to the site. This in itself may be a disadvantage, both for transport, and if variable site conditions require an unexpected alteration.

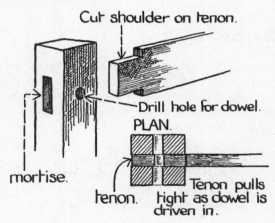

A more practical alternative is to use a rebated joint, which can be assembled on site, and replaced or altered as necessary. Although easier to cut and drill in a workshop, this can be done on site. The joint can be fixed with a coach bolt or with a length of studding fastened at each end with a washer and nut. A neat and secure finish is given if this is counterbored and tightened with a socket. Studding can be cut to length with a hacksaw.

Stiles with nailed rails are the easiest to assemble, but are not as robust or durable as jointed and bolted stiles.

Steps

The step should not rest on a rail, or a see-saw action can develop as the step supports settle into the ground with use. Allow a gap of at least 50mm between the step and the rail beneath it. The step should overhang its support by about 40mm, to help protect the top of the support from wear and weather. The easiest method of fixing is to drill and skew nail using two 125mm galvanised nails at each end. A more secure fixing is made by using a coach screw. Although a counterbored finish is neater, it does collect water which hastens rot. The protruding head helps give grip if the step is muddy or wet.

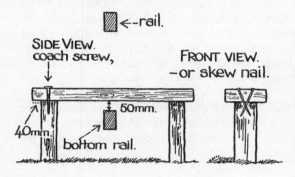

Finishing

Always weather the top of the posts so that water runs off quickly, and chamfer to give a smooth finish for the hand. A surform is the easiest tool to use. For those with the tools and time available, attractive handholds can be made; a satisfying job if you consider the number of times it will be appreciated over the years by those who climb the stile.

The top rail and step should also be chamfered for a perfect finish.

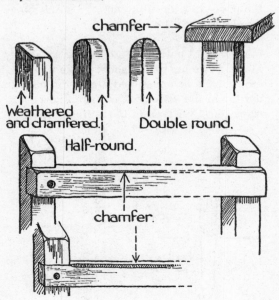

HOLE DIGGING

There are various ways of digging holes for stile posts and step supports, depending on the type of ground.

a   Post hole borer. This is the most efficient tool to use in loam or clay-loam soils without stones or roots.

b   There are several types of spade suitable for hole digging. The grafter is a long thin spade with a curve at the bottom for removing debris. The Devon shovel has a pointed blade, and a long, often curved handle.

c   Crowbar. This is often needed to break up stony ground or lever out rocks.

d   Shuv-holer. This is useful in loose sandy soil, or for removing debris from a large hole dug with a spade, but it is rather a heavy item of equipment to carry far.

See chapter 5 and the suppliers' names and addresses (p181) for further details.

Try and dig the hole neatly, with vertical sides. It should be of a diameter just large enough to use a punner (see p43) to compact the spoil around the post. Pile the debris up on an opened-up fertiliser bag, so that it is easy to shovel up and re-use. A hole of 750mm depth is normally needed for the stile post, and 500mm for the step supports.

Put the post in position, and ram soil around the base with a punner. Carefully fill up the sides with soil, ramming securely in layers about 50mm deep. If the ground is too rocky to fix the post securely, concrete must be used (see below and page 186).

Apart from the standard tools described above, various improvised tools have been made for stile construction, which may give some useful ideas.

a   Scaffold pole with old pick-ended mattock head jammed in one end, and useful for cutting down the sides of holes to keep them straight.

scaffold pole.   pick end.

b   Punner, made of plumbing pipe, with T or elbow section at one end.

c   Cut-down trowel, old Marvel tins etc for removing debris from narrow holes.

d   Plastic scoops (see p43) for removing water from holes.

Concrete

This is best used as a dry mix, as sufficient water will be absorbed from the ground to activate the cement. If a wet mix is needed, perhaps for use near the surface of the ground, the easiest method is to take the cement and aggregate to the site in a fertiliser sack, add the water, and roll the bag around on the ground until thoroughly mixed.

A mix of one part cement to $4\frac{1}{2}$ parts all-in aggregate (maximum size 20mm) is suitable. If mixed wet, a maximum $\frac{3}{4}$ part water will be sufficient. Try to keep the stile out of use for a few days, if at all possible.

GENERAL PROCEDURE

This describes the procedure for erecting a wooden step-over stile, taken to the site in kit form, and bolted together through rebated joints. This is probably the best method of stile building, as it combines the efficiency of workshop production with relative ease of transport.

1   Dig holes for stile posts to the required depth.

2   Lay posts and rails on the ground, assemble with coach bolts or studding, but do not fully tighten nuts.

3   Set posts into position, check uprights and rails with spirit level, and tighten nuts.

4   Ram soil around posts using punner or other suitable tool, compacting it in layers.

5   Dig holes for step supports, positioning them so that the step projects about 40mm beyond the support.

6   Set step supports in position. Level, using step and spirit level. Ram securely.

7   Attach step to supports using nails or coach screw.

8   Chamfer step and top rail. Weather and chamfer handhold.

9   Attach wire or fill wall gap to make the boundary stock-proof.

10   Lay surfacing material on the ground where people step off the stile to prevent muddiness and erosion.

POST DRIVING

Some groups prefer to construct stiles by the simpler method of knocking the stile posts and step supports into the ground. This depends on the factors discussed on page 148, and on the cost of the materials being used. A hardwood or treated softwood stile should always be positioned as described above, whereas a stile of split chestnut, costing perhaps a fifth of the price and with an estimated life of only 10 years without repair, can well be knocked into the ground. Kent County Council have erected many stiles in this way (see p154), and have found that their cheapness and speed of construction outweigh the disadvantage of lower durability.

The Chiltern Society use a post hole borer for the stile posts, but knock the step supports in with a mell or post-driver (upside down), as this saves time.

Note the following points concerning post-driving:

a   Do not weather the post until after it is driven in, or it will be difficult to get the post straight. Any damage to the post top can then be sawn off.

b   Roughly chamfer the top of the post before hammering it in, as this lessens the chance of it splitting.

c   Use a mell or a Drive-all. If there is nothing other than a sledgehammer available, protect the post top with a tin can or a small piece of wood nailed to the top.

d   Make a pilot hole with a crowbar or short stake. The top of the short stake thus takes most of the damage, and starts the hole so the top of the stile post is easier to reach when this is driven in to its final depth.

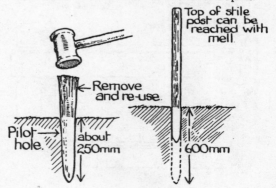

Stile Designs

British Standard

This is based on British Standard 5709:1979, amended 1982.

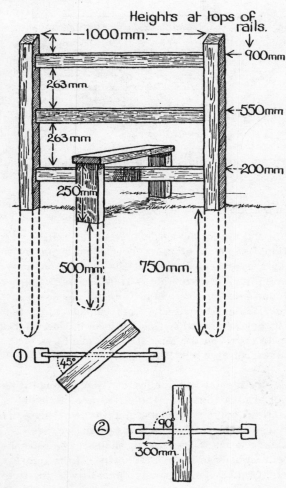

Timber:

Stile posts	2 of 100 x 100 x 1750
Rails	3 of 75 x 50 x 1100
Step	1 of 175 x 50 x 900
Step supports	2 of 150 x 75 x 750

The rails are stub-mortised full size into the posts for a depth of 50mm. The step should be set at an angle of between 45 and 90 degrees to the rail. The choice depends on the slope of the ground and the preference of the stile builder. (1) above is comfortable for most people. Less agile walkers may prefer (2), which allows one to face sideways while crossing the stile, holding onto the handhold throughout.

Rebated stile

This example is shown with two steps, for use on a slope, and a handhold to aid the less agile. The rails are fixed using the rebated joint described on page 150.

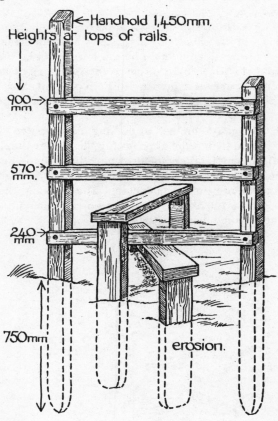

Timber:

Stile posts	1 of 100 x 100 x 1750
	1 of 100 x 100 x 2200
Rails	3 of 75 x 50 x 1200
Steps	2 of 175 x 50 x 900
Step supports	2 of 150 x 75 x 750
	1 of 150 x 75 x 500
	1 of 150 x 75 x 1000

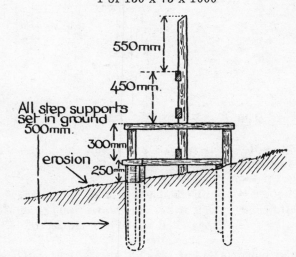

The heights of the steps and rails are given as an example only, and will have to adapted to fit the slope. The major problem is choosing the spacing of the rails so they do not coincide with the steps. It may be better to cut the rebates on site, which

can be done after the post holes have been dug and
the positions tried out. The gap between the rails
should not be more than 300mm, and a rail with
its top edge lower than 200mm is wasted. As
shown in the example above, the lower step has a
rise of less than 300mm on the lower side of the
slope, to allow for erosion to expose more of the
step support. Surfacing will help prevent this. The
upper slope is protected by the 'ground level' step.

An alternative method of making a lower step is
to rebate it into the support for the higher step.
This is particularly useful where the ground levels
are different on either side of the stile, as often
found on old field boundaries. The step support
should be 100 x 100, and the step 100 x 50mm, for
a strong and neat construction.

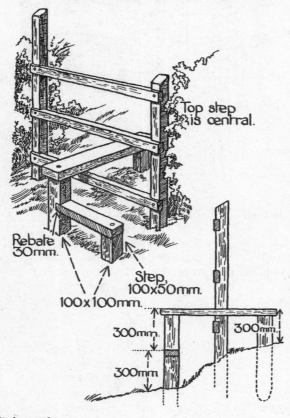

Stake stile

This stile can be quickly erected by inexperienced
volunteers using the minimum of tools.

Timber: beech or chestnut, peeled, untreated.

Stile posts	2 of 125 diameter x 1500 posts pointed at one end, dipped in creosote
Step supports	2 of 125 diameter x 1000 posts pointed at one end, dipped in creosote
Rails	3 of 125 diameter half-round x 1100
Step	1 of 125 x 50 x 1000, of cheap hardwood, or creosoted softwood

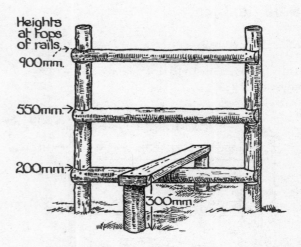

Drive posts in using Drive-all or mell. Drill all
nail holes.

Chiltern stile

This and the following design are for stiles
without separate steps. Their advantages are that
they are quick to construct and install, and use
a minimum of materials. They are useful on
ground where holes are difficult to dig, as only
two holes are required. Their disadvantage is
that they are more difficult to climb, and a lot of
strain is put on the top rail or handhold as the
walker pulls up over the stile.

The Chiltern stile is being installed in quantity by
the Chiltern Society. The stiles are made of
tanalised softwood, assembled in a workshop, and
are designed to fit in the back of an average
hatchback car. They weigh about 15kg, and take
about an hour to install. Because of the lightweight timbers used, their durability is likely to
be lower than that of other designs in this chapter.
A separate step can be added to make the stile
easier to climb.

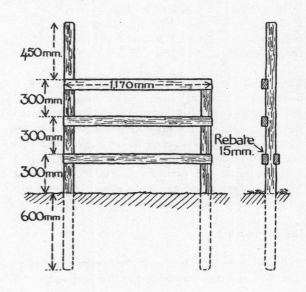

Timber:
Stile posts   1 of 75 x 75 x 1950
                 1 of 75 x 75 x 1500
Rails   3 of 75 x 38 x 1170

All rails drilled and fixed with 10mm coach bolts:
4 of 10 x 125mm coach bolts for rails
2 of 10 x 150mm coach bolts for step-rails

Wortley stile

The middle rail is mortised into the middle of
the stile posts so that the footrails project well
beyond the edges of the middle rail, thus giving
a secure foothold. The top rail is rounded and
mortised into the stile posts, with metal straps
for added security. This is a robust stile, and
comfortable to sit on as you climb over.

Never strengthen the top of a stile by putting in
a double rail, as the extra width makes the stile
awkward to straddle, and it is possible to catch
one's foot in the gap.

Timber:
Stile posts   2 of 100 x 100 x 1650
Top rail   1 of 100 x 100 x 1500
Middle rail   1 of 75 x 50 x 1130
Footrails   2 of 75 x 63 x 1400
Dowelling for mortises   12 x approx 600

Mortises:
Top rail   90 x 45 x 65 deep, spaced 1010
                 apart
Posts   65 x 45 x 65 deep, 295 from top

Fittings:
Coach bolts, nuts and washers   2 of 10 x 200
Metal straps (20 SWG) punched
for 6 x 6mm holes   2 of 40 x 900
Galvanised nails   12 of 65mm

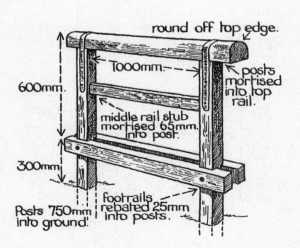

round off top edge.
1000mm.
600mm.
posts mortised into top rail.
middle rail stub mortised 65mm. into post.
300mm.
Posts 750mm into ground.
footrails rebated 25mm into posts.

STILES FOR EXISTING FENCING

The designs below are for fitting stiles to existing
fencing, without having to make a gap or re-strain
the fence.

Step for post and wire fence

This requires an existing fence post strong enough
to provide a secure fixing for the handhold.
Alternatively, a new post can be dug in. Two
steps will be needed for a fence higher than
900mm, and should be set at 90 degrees to the
fence. The steps are shorter than usual to make
them inaccessible to sheep. If the wire is
barbed, remove the barbs for a width of one
metre. This is most easily done with two pairs
of pliers.

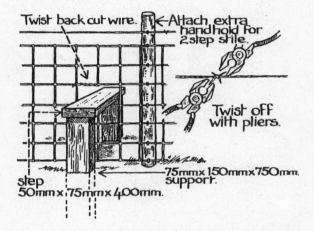

Twist back cut wire. Attach extra handhold for 2 step stile.
Twist off with pliers.
step 50mm x 75mm x 400mm.
75mm x 150mm x 750mm. support.

For sheep fencing, adapt the heights of the steps
to fit through the spaces in the fencing. It should
only be necessary to cut one vertical wire to fit
each step through. Neaten by twisting the cut
ends back into the fencing.

If the existing fence posts are not strong enough to
provide a handhold, a complete stile will have to
be fitted. The stile posts should not be used to
strain the fence. The stile rails must be fitted to
coincide with the heights of the wires. The wires
are then stapled against the rails, but without
straining the fence. The rebated stile (p153) is
suitable, or in the case where postholes are
difficult to dig because of the fence, the driven-in
stake stile may be a better choice (p154).

Stile for post and rail fence

This design is recommended by the Battleby
Display Centre (Countryside Commission for
Scotland Information Sheet 4.9.11). It should
only be used if the fence is in good enough condition to give a secure fixing for the handhold.
Note that the rise of each step is greater than
previously recommended, but is reasonable for

the height of fence being negotiated.

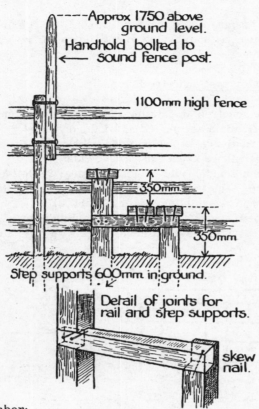

Approx 1750 above ground level.
Handhold bolted to sound fence post.

1100mm high fence

350mm.

350mm

Step supports 600mm in ground.

Detail of joints for rail and step supports.

skew nail.

Timber:
Step supports	2 of 100 x 100 x 900
	2 of 100 x 100 x 1250
Step rails	2 of 75 x 75 x 500
Steps	3 of 150 x 50 x 1000
Handhold	1 of 75 x 50 x 1000

STEP-THROUGH STILE

This stile is different in principle from the previous designs. Although the top rail is low, the narrow width means the walker cannot swing the leg sideways over the stile, but must lift the knee straight up. For this reason, this design is not suitable for paths likely to be used by the infirm. The width should not exceed the dimension shown, or the stile will not be stockproof. It is wide enough for most backpackers.

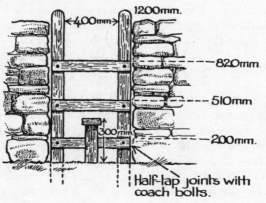

400mm. 1200mm.
820mm
510mm
300
200mm
Half-lap joints with coach bolts.

The main advantage of this design is in the saving of timber, but it is also useful in narrow wall or hedge gaps, where there is not enough room for a step-over stile.

Timber:
Stile posts	2 of 100 x 100 x 1950
Rails	3 of 87 x 30 x 600
Step supports	2 of 100 x 75 x 850
Steps	1 of 150 x 50 x 750

SQUEEZE STILES

Below are two designs of wooden squeeze stiles, which are wide enough for the average walker to squeeze through, but are proof against most stock. Design B is not suitable for fields where calves or sheep are kept. Stile A is useful at bridge ends, where it is awkward to fix a separate step (see p109).

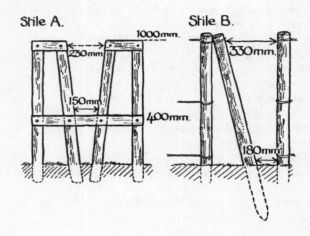

Stile A.
230mm
150mm
400mm.

Stile B.
1000mm
330mm
180mm

STILE FOR DISABLED PEOPLE

The Countryside Commission recommend the following design for disabled people (Countryside Commission, 1981, Advisory Series No 15). The steps are low and wide, and the top step is wide enough to allow a person to sit on it and swing his legs over.

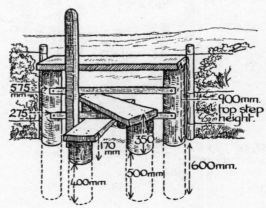

575 mm
275 mm
170 mm
350
400mm
500mm
600mm.
900mm. top step height.

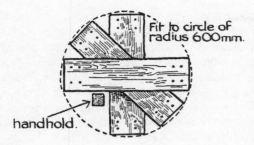

Fit to circle of radius 600mm.

handhold.

Timber:
Step supports of peeled larch or pine, of about 200mm diameter. Two each of lengths 1500, 850 and 570mm.

Steps	3 of 50 x 300 x 1200
Rails	2 of 87 x 30 x 1100
Handhold	1 of 75 x 75 x 2000

Ladder Stiles

Ladder stiles are used mainly in upland areas, to cross dry stone walls and deer fences. As well as being robust enough to stand heavy use and weather over many years, they serve as useful 'waymarks', particularly for leading people down the correct route off the fellside into enclosed farmland.

They should normally be assembled on site, both because they are too heavy to easily transport once made, and because exact dimensions are difficult to predict. These dimensions will depend on the height and width of the wall or fence, and the slope of the ground. Often hole-digging is a problem because of rocky ground, and the stile cannot go exactly where intended.

The most important feature is that the steps are securely fixed and regularly spaced on any one stile, or else it is easy to stumble. A spacing of between 300 and 350mm is suitable. The length of the platform cannot be determined until the 'A' frames are assembled, and it is better to measure and cut the platform timber on site.

Two designs are given below.

Step-ladder stile

Although this stile requires more timber and may take longer to construct that the rung-ladder stile, it is recommended as it is easier to climb.

The steps can either be rebated into the 'A' frames, or held by extra pieces of timber cut as shown, and nailed or screwed to the 'A' frame. These pieces can be easily cut on site and attached so they grip the steps firmly. They can be removed if steps need replacing. The steps should be 50mm thick to give many years of use, and spaced

to a maximum rise of 300mm. To cross a wide barrier such as a wall, a platform will need to be attached to the top steps, as in the rung-ladder stile.

The stile should be tightened with a tie rod of studding or fencing wire to hold the steps firmly in position. A strip of expanded metal stapled to each step will prevent them becoming slippery in muddy conditions.

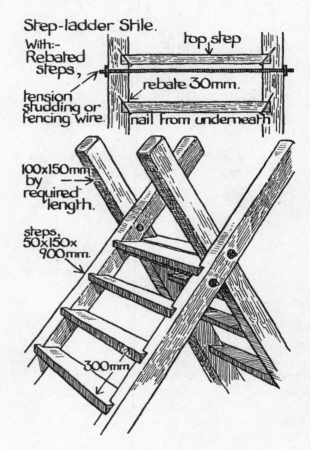

Step-ladder Stile.
With:-
Rebated steps,

top step

rebate 30mm.

tension studding or fencing wire.

nail from underneath

100x150mm by required length.

steps, 50x150x 900mm.

300mm

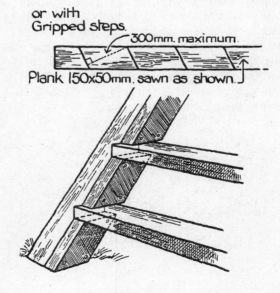

or with Gripped steps.

300mm. maximum

Plank 150x50mm. sawn as shown.

Rung-ladder stile

These are suitable for remote locations where they are unlikely to be used by less agile walkers. The rungs must always be rebated, and not simply held by nails which could become loose after a year or two. A tie rod is not required, as the rungs hold the frames in position.

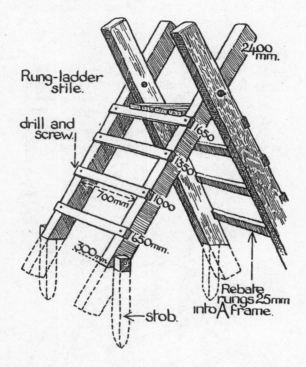

Rung-ladder stile.

drill and screw.

2400 mm.

1650

1350

700mm

1000

650mm

300mm

stob.

Rebate rungs 25mm into A frame.

Timber:
'A' frames	4 of 150 x 100 x 2400
Rungs	8 of 100 x 50 x 900
Platform	about 2500 length of 100 x 50 to be cut on site
Stobs	4 of 100 x 100 x 500

The crossover of the 'A' frames means that the platform has to be offset. The gaps on either side of the platform can be filled with shorter pieces, which looks neater and prevents the danger of a foot getting caught. However, as these can only easily be fixed by nailing to the 'A' frame from underneath, they are not very strong. For simplicity of construction, it is probably better to leave the gaps on either side.

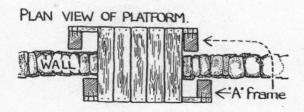

PLAN VIEW OF PLATFORM.

WALL

'A' frame

PROCEDURE

1   If there is a choice of site, look for a position where the ground is level. This not only makes the stile easier to build, but lessens the problem of the ground eroding at the base of the stile. On a slope, this can progressively expose the stobs until the stile is left suspended and liable to collapse.

If the stile is to cross a wall, avoid places where the wall curves or is irregular in width, or the 'A' frames will be difficult to line up.

2   Hold one piece of an 'A' frame roughly in place, to mark the position of the hole. Usually this will be between 400 and 500mm away from the wall. Dig the holes for one 'A' frame to a depth of about 300mm.

3   Two people then hold the pieces of the 'A' frame in place, while a third marks the line where they cross. The frame should not touch the wall.

4   Lay the frame on the ground at the marked angle, drill and fix with coach bolt. Put the head of the bolt to the inside of the stile.

5   Set the 'A' frame into position. Knock in a stob, as deep as possible, immediately next to the base of the frame. Nail the stob to the frame. The stob may have to be sawn off if rocky ground prevents it being knocked in to its full depth. Backfill with earth, ramming continuously to get it very firm.

6   Erect the second 'A' frame in the same way.

7   Mark the position of the platform so that it clears the wall. Use a spirit level to get it horizontal. Measure from this mark down to the ground, and divide by 350 to give the positions of the rungs.

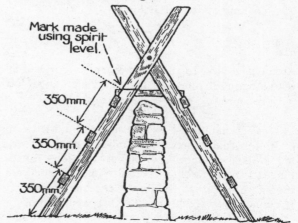

Mark made using spirit level.

350mm.

350mm.

350mm

If this leaves an awkward gap at the bottom, adjust the spacing up to a maximum of 400mm. Try and use the same spacing on either side of the stile. The bottom rung can be set lower than 350mm to allow for erosion, but preferably prevent the erosion rather than allow for it. Note that the rung supporting the platform is set about 50mm below the horizontal mark, to allow for the thickness of the platform.

8   Cut 25mm deep rebates, and screw the rungs in place. Chamfer the top rung to give a horizontal face to which the platform is nailed. Chamfer the ends of the platform timbers and tops of the 'A' frames, which form the handholds.

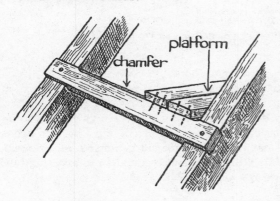

9   Surface the ground at either side of the stile to prevent erosion. Stream gravel or packed stone are suitable, laid in an informal manner that does not detract from the surroundings.

Stone Stiles

Building stiles in stone walls requires some skill and knowledge of walling techniques. These techniques are described in detail in 'Dry Stone Walling' (BTCV 1977). The information here gives the design and average measurements of various stiles, with a few notes on construction methods.

Step stiles

The steps should protrude at least 300mm from the wall. Ideally, stones should be used which are long enough to protrude either side, but if not available, the stones should be embedded for at least twice the length of the step. Long face stones help to hold the step securely. Railway sleepers can be used if there are no suitable stones available, but these are not as attractive or durable. A heavy through stone is needed to protect the top of the wall. A timber handhold

can be added; this also acts as a useful waymark if the path is not worn.

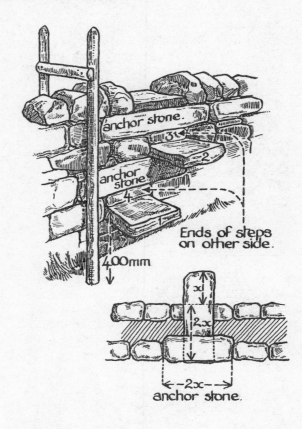

Step-over stiles

Traditional 'stone hedges' in Cornwall sometimes have stiles with hewn granite rails.

A method likely to be more practical is to use scaffold poles or any other suitable iron pipe. These weather to look reasonably unobtrusive. The poles can be fitted to from the stepped profile shown above, or as an upright stile, with a separate step if necessary.

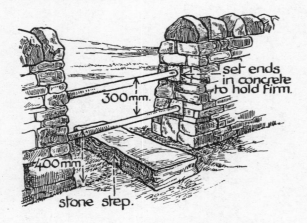

set ends in concrete to hold firm.

300mm.

400mm.

stone step.

Squeeze stiles

There are many variations of the squeeze stile described in 'Dry Stone Walling', depending on the available stone. The principle is that the average person can squeeze through the gap, but stock cannot. An extra barrier may need to be added at lambing time.

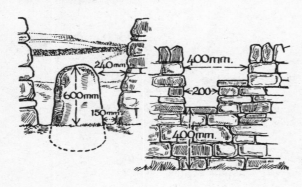

240mm

400mm.

600mm

<200>

150mm

400mm.

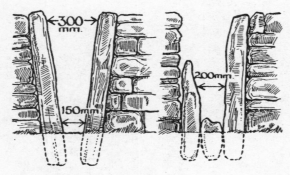

<300mm>

200mm

150mm

Rung stiles

These are traditional in Cornwall and the Isles of Scilly, and work on the same principle as a cattle grid. The width of the gaps and the number of rungs depends on the stone available, but usually four or five rungs are used. A large stone can be used for the middle rung, so that it sticks up as a sill about 300mm high. The gaps

between the rungs have to cleaned out from time to time for the stile to be stockproof.

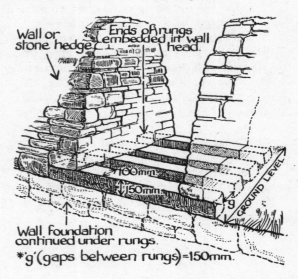

Wall or stone hedge

Ends of rungs embedded in wall head.

100mm

150mm

GROUND LEVEL

9

Wall foundation continued under rungs.

*'g'(gaps between rungs)=150mm.

Oddities

These are a few of the countless variations to be found in the structures people have built to cross boundaries. They are not generally recommended for all situations, but can add an interesting landmark to a walk.

Clapper or hammerhead stile

This is probably a design from the 18th or 19th Century, as it would have been ideal for use by ladies in long skirts. There were two examples of such stiles at Folkestone, Kent, and in a churchyard at Hungerford, Berkshire, which were called clapper stiles by a writer in 'Country Life' magazine. There is also one at Sissinghurst Castle, Kent. A modern equivalent has been built by K H C Trodd of the Forestry Commission at Sidwood Picnic Place, near Kielder, Northumberland, and his design is shown below. The stile has operated for 6 years without requiring any maintenance, but was built of untreated spruce which has a limited life.

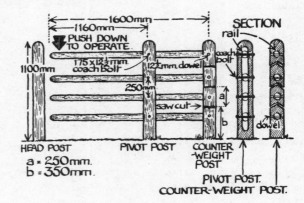

1600mm

1160mm

PUSH DOWN TO OPERATE

SECTION

rail

1100mm

175x12½mm coach bolt

12½mm dowel

coach bolt

250

saw cut

a

b

dowel

HEAD POST

a = 250mm.
b = 350mm.

PIVOT POST

COUNTER WEIGHT POST

PIVOT POST.
COUNTER-WEIGHT POST.

Timber:

Posts	2 of 150 diameter x 1700
	Pivot post: cut slot 55 x 900
Counterweight	1 of 150 diameter x 1100, cut
	into three sections of 250 and one
	of 350mm length
Rails	4 of 55 diameter x 1600

Fixings:

Coach bolts	4 of 12½ x 175
Dowel	12½ x 600

Stair stiles

There are many examples of these, both in wood and stone. Stone stairs are particularly associated with churchyard walls.

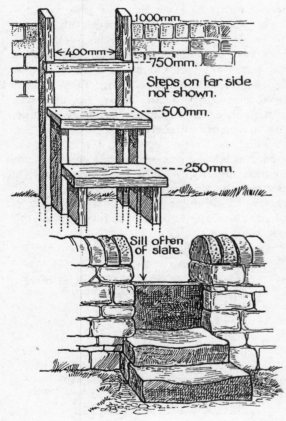

Timber:

Stile posts	2 of 75 x 75 x 1600
Rail	1 of 75 x 32 x 550
Steps	4 of 175 x 50 x 450
Step supports	4 of 100 x 50 x 700
	4 of 100 x 50 x 950

Apple box stile

The design was seen, not surprisingly, at a fruit farm; Kings Coppice Farm, Cookham Dean, Berkshire. The sides of the boxes could be removed to improve the appearance of the stile, in effect using the boxes only as formwork for the concrete.

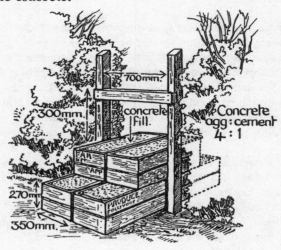

Barbed-wire covering

As a quick improvement to an existing stile, barbed wire can be covered either with a section of hosepipe, cut spirally and slipped over, or with a fertiliser bag folded over and tied with string. The latter, if of a bright colour, also acts as a waymark.

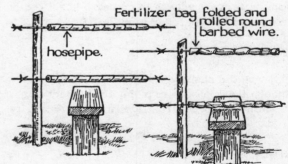

Dog latches

These are useful on popular dog walks.

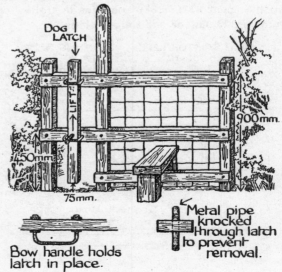

161

Gates

A type of gate often seen on public paths is the kissing gate, which cannot be left 'open' and accessible to stock. It also prevents access to motor cyclists and bicyclists, although lightweight cycles can be lifted over. The disadvantage of the standard kissing gate is that it is inaccessible to wheelchairs, and awkward for a walker with a bulky rucksack.

Less commonly seen is the bridle gate, which is about 1500mm wide, and specifically designed for use on bridlepaths. Many footpaths and bridle-paths pass through field gates, which can be between 2½ and 3½ metres wide, and are notorious for being badly hung or not hung at all, and tied up with baler twine. Such gates are awkward for both landowner and walker or rider, and may be the cause of conflict. An unworkable gate is liable to be either left open, or broken whilst being negotiated, or is avoided altogether with resulting trespass and possible damage.

This section gives advice on the construction and hanging of kissing and bridle gates, and on adaptations to prevent access by motor bikes. Gates are usually bought ready made (p182 12.1-12.2).

Responsibility

Gates are the responsibility of the landowner, although those on public paths are subject to a 25% minimum grant from the Highway Authority.

BRIDLE GATES

The main requirements are as follows:

Dimensions

British Standard 5709:1979 specifies the following minimum dimensions and timber sizes.

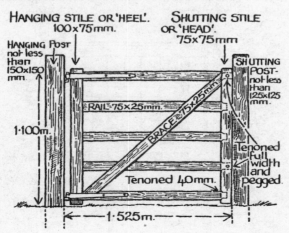

Materials

Bridle gates can be either wooden or metal, but wooden gates are usually preferred for their durability and appearance. See page 183 for details of suitable types of wood.

Construction

The gate must be braced with a diagonal member, and the top rail and the rail second from the bottom should be mortised the full width of the uprights (called 'stiles'), and pegged.

Hanging

The hanging and shutting posts should be embedded at least 1100mm. The gate should normally be hung so that it swings shut. In order to make use easier for horseriders, hinges have been devised which allow the gate to swing both ways (see Countryside Commission Management and Design Note 5). These have not proved to be entirely satisfactory, as the parts are difficult to obtain and require very precise fixing to operate properly.

In order to make a one-way gate swing shut, the hooks are slightly offset on the hanging post, the upper hook being about 30mm offset in the direction in which the gate closes. The bottom hook should project about 5mm further than the upper hook.

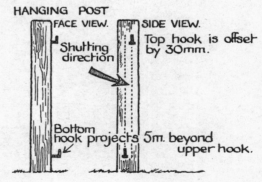

In areas where vandalism may be a problem, it is a good idea to invert the top hook to prevent the gate being lifted off. Alternatively, if the gate has already been hung and it is not possible to turn the top hook through 180 degrees, a metal peg can be driven in above one or other of the hooks. The latter is not so good if the gate needs to be removed for repair or replacement.

Latching

A good latch should be:

a Self latching when the gate swings closed.

b   Proof against an ingenious animal.

c   Designed to function for a long period, even if the gate 'drops' with time and use. This is more important with gates wider, and so heavier, than the standard 1.5m bridle gate.

Different types of latches are shown below.

Spring latch

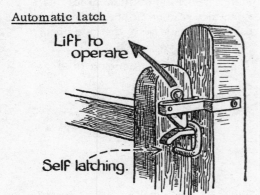

Advantages.
Easy to operate from either side. Still works if gate drops.

Disadvantages.
Awkward to fit so that the tension is enough to keep the gate closed without being difficult to operate. Action of using lever puts strain on the hinges. Not self-latching.

Automatic latch

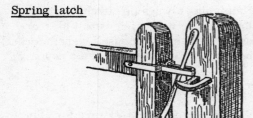

Advantages.
Easy to fit and durable as long as the gate does not drop. Self-latching.

Disadvantages.
Awkward to operate from far side.

Loop latch

Advantages.
Easy to fit and to operate. Durable if of rigid metal or chain. Works even if gate drops slightly.

Disadvantages.
Can be worked by animals. Frequently improvised with wire or baler twine, which are not durable and not respected by users.

Pivot latch

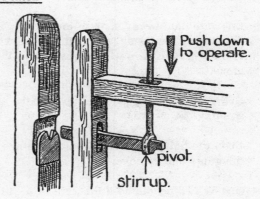

Advantages.
Easy for rider to operate from either side.

Disadvantages.
Not very robust, and can easily be bent out of alignment. Fitting requires making a large hole in the top rail for the stirrup to pass through, resulting in a slack fitting for the rod.

During the Brecon Beacons Pony Trekking Project (Bryant, 1978), several different types of latches were used. Automatic latches were found to be the most reliable and effective.

Positioning

The gate should always be positioned so there is room for the horse to stand to one side while the rider leans over and opens the gate. The gate should normally open away from the road or track, and onto the field or bridleway.

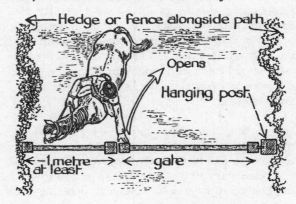

1   Dig hole for hanging post, at least 1100mm deep.

2   The lower hook is fixed first. Take measurement from the gate, allowing for ground clearance. Drill a hole of a diameter slightly less than that of the hook. Drive in the hook with a lump hammer.

3   Lift gate onto lower hook to locate the position of the upper hook. If the latter is to be inverted, this will be just above the hinge.

4   Drill hole in post, and fix upper hook, pointing downwards.

5   Undo the bolts attaching the upper hinge to the gate, and remove.

6   One or two people hold the gate in position, while the upper hinge is slotted onto the hook, and bolted back onto the gate.

7   The gate can now be used to locate the position of the shutting post (or posts in the case of kissing gates).

Kissing Gates and Barriers

Kissing gates were traditionally built to be stock proof with the gate in any position, while allowing the passage of people. Their advantage was that they were virtually foolproof, and did not need latches or springs to be stock proof, nor could they be accidentally left open, or purposely propped or tied open.

For modern day use they have a particular disadvantage, as they are not passable by walkers with large rucksacks, and if made large enough to be passable, are no longer sheep proof. Before deciding on a kissing gate, assess carefully the use of the site, as it may be just as effective to have an ordinary gate with a spring and latch, so that it is self-closing. A kissing gate that needs a spring and latch to be stock proof has no advantage over an ordinary gate, except that it stops motor cycles.

The following basic dimensions are given for various uses. These are not foolproof, as sheep vary in size and in their tendency to escape, just as walkers vary in their stoutness and in their competence at packing a rucksack. Kissing gates are not lamb proof, and extra wire fencing and a

temporary latch may be needed at lambing time.

Except in the Countryside Commission for Scotland design shown below, the width of the gate itself makes no difference to the accessibility of the gateway. For economy of materials and ease of hanging, a 700mm wide gate is sufficient. The significant dimension is the area of the space made by the guard rails with the gate in mid-position. Rails must always be attached to the outside of the posts so that stock cannot loosen them by leaning on them. Dimensions are internal, measured between the inner edges of the posts.

Ground plans:

a   This is sheep proof, but awkward for stout walkers and those with rucksacks.

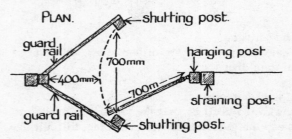

b   The design below is not stock proof, as a sheep can turn around in the space between the guard rails and 'nose' the gate open.

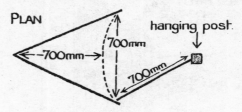

c   BS 5709:1979, amended 30.9.82, specifies the use of three guard rails, making a space 500mm square when the gate is in mid-position. This design is not recommended, as it is a squeeze even for an unloaded walker, and impassable for a walker with only a small rucksack.

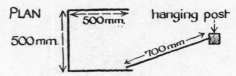

d   The following design is by the Countryside Commission for Scotland (see CCS Information Sheet 4.8.4). It is simple to construct as it is based on a square, and so the joints between rail and post are easy to make. However, in order to be passable to a walker, the gate has to be very wide, which uses more materials than a traditional plan of kissing gate.

PLAN.

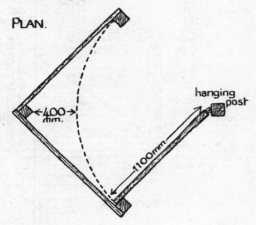

e   The design below, although more complicated, is recommended. It is both sheep proof and passable by walkers with large, bulky rucksacks. It does not require a spring or latch to be stock proof.

PLAN.

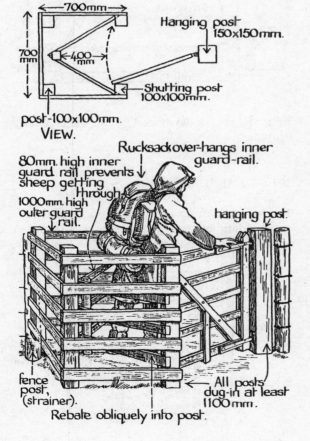

VIEW.

Procedure:

1   Position hanging post, and hang gate (see above).

2   Position shutting posts.

3   Position post for sheep guard rail. Attach rails.

4   Position remaining two posts, and attach full height rails.

5   Position straining posts and restore fence.

6   The space between the two guard rails will not be grazed or trampled, and will be a nuisance to walkers if it becomes overgrown. Use a herbicide such as glyphosate (see p57) to prevent this happening.

Stileway

These gateways are not stock proof, but are useful for preventing motor bike access.

PLAN

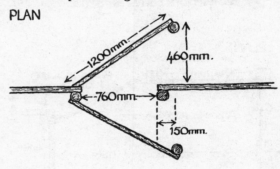

The Forestry Commission at Westonbirt Arboretum, Gloucestershire, have used a similar method for making a temporary stileway through a post and rail fence. As the path becomes worn, the original rails can be replaced and the stileway moved to another position.

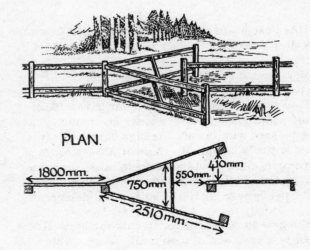

PLAN.

165

Stileway for wheelchairs

This stileway allows access for pedestrians, pushchairs and wheelchairs, but prohibits horses and motor cycles. The heights of the rails give maximum clearance for the elbows and feet of wheelchair users. The U turn prevents motor cycle access, and the rail excludes horses.

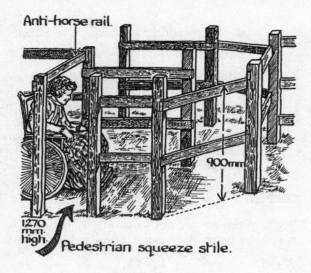

Anti-horse rail.

900mm

1270 mm. high.

Pedestrian squeeze stile.

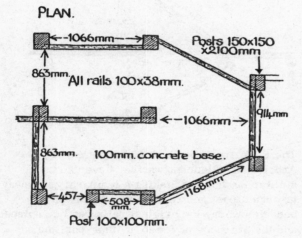

PLAN.

←1066mm→ Posts 150x150 x2100mm

863mm All rails 100x38mm.

←1066mm→ 914mm

863mm. 100mm. concrete base.

1168mm

←457→ ←508 mm.→

Post 100x100mm.

This design was developed by the Upper Lea Valley Area Countryside Management Project.

Motor cycle barrier

This gateway provides access for horseriders and pedestrians, whilst access for motor cyclists is blocked with the 300mm high barriers. It is not entirely bikeproof, as determined groups of motor cyclists could manhandle bikes over the barriers, while one person holds the gate open. It has however proved to be a useful deterrent.

The gate is not particularly easy for horseriders, but is a good test of their proficiency. This

gateway can be installed alongside the one shown above to allow wheelchair access.

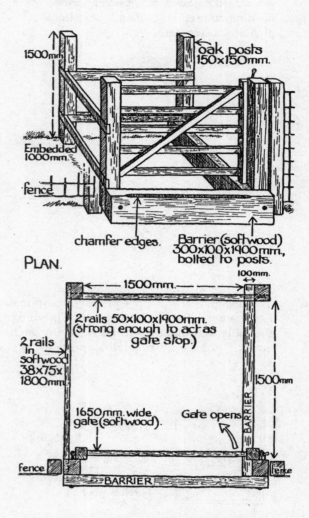

1500mm oak posts 150x150mm.

Embedded 1000mm.

fence

chamfer edges.

Barrier (softwood) 300x100x1900 mm, bolted to posts.

PLAN.

←————1500mm.————→ 100mm.

2 rails 50x100x1900mm. (strong enough to act as gate stop.)

2 rails in softwood 38x75x 1800mm

1500mm

1650 mm. wide gate (softwood). Gate opens

BARRIER

fence

BARRIER fence

This design was developed by the Tame Valley Project.

HORSE BARRIERS

Barriers at entrances to footpaths to prevent horse riding are fairly simple to construct. The usual type is a staggered bar, as shown below.

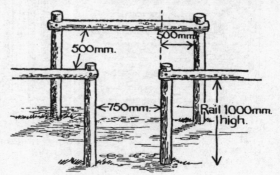

500mm

500mm.

←750mm.→ Rail 1000mm. high.

Bridlepaths which receive conflicting use by horseriders and walkers can be divided by a barrier. This prevents the entire surface becoming churned by horses, and lessens the danger of collision between horses and walkers.

However, as with any fences that enclose a path, a barrier will add to the amount of maintenance required. In any unshaded situation, grass will grow along the line of the barrier, and possibly block the footpath if use is not sufficient to keep it trampled. The walker can then be faced with a choice of either a very muddy or an overgrown path. In this sort of situation, use of a total weedkiller such as glyphosate may be justified.

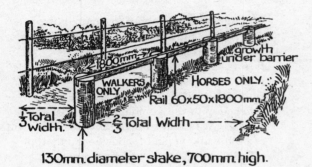

A simple but effective barrier can be made by knocking stakes in, about 600mm high and 1000mm apart. Horses are restricted to one side only, but walkers can walk either side, retreating to the safety of the footpath if horses approach. As well as being cheaper in materials, there is less likely to be the problem of growth described above, as there will be some trampling between posts. The posts should have flat tops to lessen the chance of injury to horse or rider.

This system has been used successfully at Frensham Ponds, Surrey.

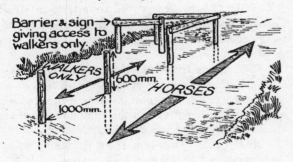

13 Waymarking

This chapter is concerned with the waymarking and signposting of paths. Waymarking usually refers to the marking of existing objects along a path to indicate the route, and signposting to the provision of separate signposts, particularly at path junctions. The technique and organisation of waymarking is discussed, and the construction and erection of simple signposts. The many techniques for producing signs are not discussed, as comprehensive information is available in other publications (see Brown, 1974, Countryside Commission, 1981 and Countryside Commission for Scotland Information Sheets).

Principles of Waymarking

The Waymarking Study Group of the Countryside Commission, set up in 1973, defined waymarking as follows:

"To waymark a public right of way is to mark the course of the route at points along it. Waymarking is complementary to signposting, which is normally reserved for the points where a path makes a junction with a road. Signposts advertise a path in its initial direction; waymarks enable users to follow the path accurately at points where they might otherwise have difficulty."

The term 'waymark' is a very ancient one, from Jeremiah 31 v 21, "Set thee up waymarks, make thee high heaps: set thine heart towards the highway, even the way which thou wentest".

ADVANTAGES OF WAYMARKING

1   It allows walkers who are unfamiliar with an area to explore the paths without going to the expense of purchasing maps.

2   It aids route finding in woodland and scrub where even the 1:25,000 maps do not give sufficient detail.

3   It reduces accidental trespass, and allows intentional trespassers to be fairly reprimanded.

4   It clarifies the status of paths for walkers and riders.

5   It encourages the use of little known paths, both by visitors and local residents, and so reduces the frequency of clearance needed.

Note the following:

a   The general rule is to place waymarks only where they are necessary for a stranger to follow the route of the path. However, 'confirmatory' waymarks may be necessary in places where there are several clearly trodden paths, not all of which are rights of way.

b   Waymarking should only be done on paths which are clear of overhanging vegetation, and which have the necessary stiles and bridges. Otherwise, trespass is likely to increase.

The same should apply to signposting of paths from metalled roads, which is the responsibility of the local authority, but this is frequently not the case.

c   Only public rights of way should be way-marked with the standard blue or yellow arrow. It is most important, for the rights of both walkers and landowners, that un-official permissive paths or diversions are not waymarked. From the walker's point of view, the diversion or permissive path may be a less commodious route than the official path, and by encouraging use of the former through waymarking, the latter may become disused and be subject to a successful closure order.

Conversely, the result could be unsatisfactory for the landowner, if the unofficial waymarked route became official by 'presumed dedication', whilst the original route was not closed. The landowner would then have two official paths instead of one. If a landowner waymarks an unofficial diversion, in an attempt to lessen use of the official path, this could constitute mislead-ing notices, which he could be required to remove.

There are frequent cases where it is not possible to complete the waymarking of a path, either because there is a discrepancy between the route on the map and the route on the ground, or the landowner disputes the route. This is a great problem, as to partly waymark a path is confusing, but without any, the path is likely to further decline in use. The only solution, although likely to be a lengthy one, is to press the local authority to resolve the matter, it being their responsibility. This must be done either by reinstating the original route, or by a diversion order. Alternatively, a permissive or licensed path may be made as an interim measure.

Official permissive paths or licensed paths (see p13) will normally have a separate sign stating that the path is used with the owner's permission and is not a public right of way, and thus waymarking and ensuing use will not affect its legal status. Some form of waymarking is even more important than on public rights of way, as permissive paths are not shown on maps. Groups should check with the local authority to find out what scheme, if any, they have adopted for waymarking official permissive paths.

d It is not usually advisable to have a sign explaining the waymarking at the entry to a waymarked path, as this only clutters the signpost and attracts vandalism, although it may be necessary if a different system is used for a permissive path. If an arrow of the type marking the path is included on the initial signpost, this should be sufficient explanation.

Any posters describing the waymarking system are better displayed at car-parks, on parish notice boards, and at libraries, camp-sites and so on.

Waymarking can vary in form:

1 The simplest type is a non-directional mark which attracts the eye, and from which the next mark is visible. These include cairns used in some mountain and moorland areas (see p175) and 'blazes' as used on trails in America and elsewhere.

2 The next level of waymark indicates the direction of the path. A white arrow, for example, is used by the Chiltern Society.

3 The standard system recommended by the Countryside Commission indicates both the direction and the status of the path. A yellow arrow indicates a footpath, and a blue arrow a bridlepath. An additional symbol may be used to indicate a particular recreational route or long distance route.

Note that waymarking must be done from both directions along the path.

Responsibility

It is the statutory responsibility of the local authority (see p11) in England and Wales to erect

signposts at junctions of footpaths and bridlepaths with metalled roads. It is also a duty of the local authority to erect and maintain signposts and signs along the route of a path if it is considered necessary in order that strangers can follow the route of the path (Countryside Act 1968 section 27).

As waymarks are usually attached to or painted on objects which are private property (trees, fences, walls etc), the landowner's consent must always be gained first. The local authority is empowered to erect signposts in the soil of the path without the permission of the landowner (although the landowner must be 'consulted'), but it is obviously most unsatisfactory if such measures have to be taken without the landowner's agreement.

Requesting Permissions

Waymarking may appear a simple task; very suitable for local voluntary groups as practical problems are few, materials are cheap and easy to transport, and the work is satisfying for the walker. It would seem that a national campaign could quickly mobilise local volunteers into way-marking the entire public path system. However, there is more to waymarking than simply using a little skill with a paint brush, as consents must be sought and the path be made passable before any waymarks are done. It is just those areas where volunteers may feel they can give benefit by doing waymarking, that the request to the land-owner is likely to expose many unresolved problems over disputed routes, missing stiles and footbridges. This may then lead, as it did in the Chilterns, to volunteers undertaking practical work, and pressing local authorities to resolve disputes. In other areas with less active voluntary groups the problems may be too daunting and the initial enthusiasm is dampened.

Between 1976-79, the Ramblers' Association and the Countryside Commission sponsored a way-marking project, which covered eight different areas in southern England (Ansell, 1979). The waymarking officer sought the necessary permis-sions from local authorities and landowners, and organised local volunteers. To quote the project report, "The task of seeking consents from land-owners is the most daunting aspect of the way-marking work, and which deters many willing volunteers". This task is of course common to all types of work on rights of way, but these often require the closer involvement of the local authority, who then have the onus of making the initial approach to the landowner.

In summary, waymarking cannot be done in isolation, but must be part of a scheme to improve the standard of paths both for walkers and riders, and for landowners. Those who approach landowners must appreciate their problems, and in particular those of damage by trespassers and uncontrolled dogs.

The waymarking project report recommended that the following contacts should be made by a voluntary group wishing to do waymarking:

1   The local authority. As detailed on page 11 this may be either the County Council or District Council, and public paths may be dealt with by any one of several departments. Their permission is legally required, and they may already have a scheme in preparation, or be able to assist with materials.

2   Local branch of the National Farmers Union. For a scheme involving several landowners, it was found that the support of the NFU greatly eased the initial approach to landowners. In some areas they assisted by publicising the advantages of waymarking in local newsletters. The Country Landowners Association was another body whose support aided the project.

3   The Ramblers' Association. Many groups will already have waymarking well established in their area, and have the motivation to carry out the work efficiently.

4   Parish Councils. Some will have a footpath representative who takes responsibility for all matters concerning public paths in the parish, and if there is good rapport with landowners, permission to waymark may be arranged with no trouble.

It was not recommended to request permission by letter, and then assume this was granted if no reply was received. A simple form can be used to speed the process of gaining permission, combined with a visit to the farm to discuss the scheme. Some local authorities issue forms similar to the one shown below:

'I, ...of ...., being the owner/occupier of land crossed by public path(s) no(s) .... will raise no objection to members of the .... placing and maintaining waymark symbols at places on that land agreed with me, subject to the .... County Council indemnifying me against claims which may arise from the erection and existence of the waymarks'.

Signed.................Date ....

Priorities

The aim should be to have all paths maintained and waymarked. However, in most areas it will not be possible to get all the work done at once, and through routes, linking routes, and those without any major practical or legal problems will be done first. Landowners may agree to have some paths waymarked, but not others. It is usually better to do those agreed, and hope that the result will encourage the owner to let the remainder be waymarked, but any problems should be referred to the local authority.

The Waymark Arrow

Painted or fastened?

Many different methods have been used for printing an arrow on plastic, metal or small pieces of wood, and then attaching these to various objects along the path. The arrows are often printed on a circular base, so they can be placed at any angle as required. These have the advantage of being uniform and quick to put up. They are particularly suitable for use by people employed in path management, as waymarks can then easily be put up as necessary, while other work is being done, without the rather fiddly business of coping with paint and brushes.

However, it has been found that the metal waymarks, although resistant to weather, offer least resistance to souvenir hunters. Plastic signs tend to crack and peel away after a short time. For simplicity and durability, waymarks painted directly onto objects in the field are preferable.

Colour and shape

The Countryside Commission adopted the following colour system:

Footpaths   Yellow   BS range 4800 08 E 51
Bridlepaths Blue   BS range 4800 20 E 51

No colour has yet been allocated to Byways open to all traffic.

This choice has received a certain amount of criticism, as blue is difficult to see in poor light, or when the arrow is faded or discoloured. However, the system is generally accepted, and is now widely used.

The Countryside Commission arrow is quite distinctive and is unlikely to be confused with any other sort of mark or arrow. It is also a neat shape, easy to fit onto narrow stile posts and small signs, and fairly easy for even a not too steady hand to paint.

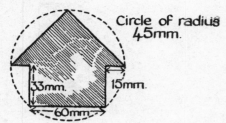

Circle of radius 45mm.

33mm. 15mm. 60mm.

The arrow was designed to appear from a distance as a blob which caught the eye, with the direction becoming apparent as the walker approached the mark.

Use

The arrows are nearly always used in the same way as traffic signs, so that a vertical arrow means straight ahead, and a horizontal arrow indicates a change of direction through 90 degrees. However, a horizontal arrow may sometimes be used as a 'confirmatory' arrow, meaning continue in the same direction, when there is no suitable surface on which to paint a vertical arrow (see example on p172).

To test the correctness of the angle, imagine the arrow tilted into the horizontal plane. Occasionally arrows are painted on a horizontal surface, but they are then not visible from a distance, and weather more quickly.

Straight on.   Turn right.

Bear right.

A criticism of the waymark arrow is that it is inflexible, and no use for indicating where a path curves around the end of a fence or wall.

Although not part of the recommended system, some waymarkers now extend the shaft of the arrow in order that a clearer picture of the route can be made. This would seem to be a sensible idea, if used with care. In the two situations illustrated below and on the following page, examples (a) show how the standard arrows can mislead, and examples (b) show the suggested use of the bent arrow.

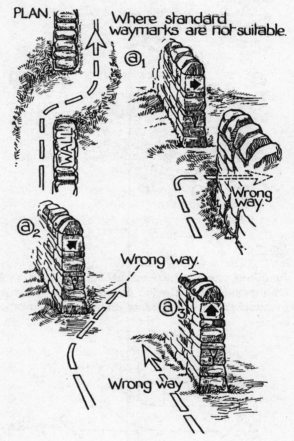

PLAN.   Where standard waymarks are not suitable.

ⓐ₁

WALL

Wrong way.

ⓐ₂   Wrong way.

ⓐ₃   Wrong way

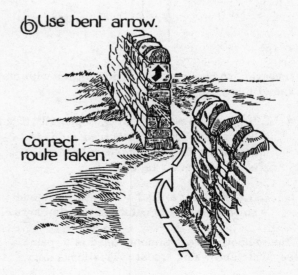

ⓑ Use bent arrow.

Correct route taken.

PLAN
Where should the waymark be placed?

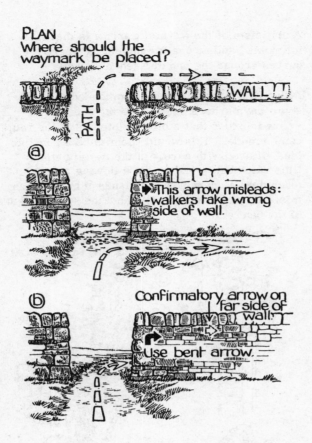

ⓐ

This arrow misleads: -walkers take wrong side of wall.

ⓑ

Confirmatory arrow on far side of wall.

Use bent arrow.

Junctions

Junctions can also present problems. The Countryside Commission recommend using a junction arrow of the basic dimensions shown below.

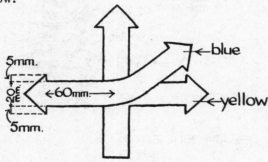

However, there are various difficulties with the use of this:

a   It is complicated to paint, especially if both blue and yellow are required and it is necessary to wait for one colour to dry before the other is applied.

b   It is difficult to find a surface large and smooth enough to paint the junction arrow.

The method more commonly used is to paint separate arrows on a post. This looks very

cluttered and confusing, and if the post is turned or broken, the possible permutations become something of a problem.

West Sussex County Council, who maintain their public path system to a very high standard, solve the problem by using finger posts at all path junctions. These are rather costly and prone to vandalism, but are attractive and easy to follow. Other ideas for junctions are given on page 178.

PAINT

Water-based emulsion paint is recommended, as it is easy to work with and to store, and can be applied to a slightly damp surface if necessary. Mistakes can be corrected if they are wiped out immediately, but the paint dries quickly enough for the second coat to be applied a short time after the first coat. Its disadvantage is that emulsion paint is not very durable, especially in exposed coastal locations. Emulsion waymarks need repainting every two to three years.

Gloss paint is more durable and gives a brighter finish, but it is awkward to use in the field because of the necessity of carrying thinners to clean the brush. It also requires a separate undercoat, so that more than one trip is necessary. Gloss paint should be used on signs produced in a workshop, or on objects 'in the field' in coastal locations.

The recommended colours are available as follows:

Yellow	Blue	
BS 4800 08 E 51	BS 4800 20 E 51	
Marigold	Forget me not	Manders
Saffron yellow	Electra	Johnstones
Tivoli gold	True blue	Crown
Goldcup	Cornflower	Dulux

Emulsion paint should be 'vinyl matt' which gives better durability than 'vinyl silk'. Sizes are as follows:

Yellow Emulsion	Gloss	Blue Emulsion	Gloss	
2½ 5	½ 1 2½ 5	2½ 5	½ 1 2½ 5	(M)
1 2½ 5	¼ ½ 1 2½ 5	1 2½ 5	1 2½ 5	(J)
1 2½ 5	½ 1 2½ 5	1 2½ 5	½ 1 2½ 5	(C)
2½ 5	½ 1 2½ 5	2½ 5	½ 1 2½ 5	(D)

Size in litres. (M) Manders; (J) Johnstones; (C) Crown; (D) Dulux.

All except the Dulux emulsion are available at retail shops, although ordering may be necessary. Dulux emulsion in both colours is only available at trade centres. Dulux gloss contains silthane which is claimed to give a tougher finish than the normal polyurethane gloss paints. In very exposed situations, Interlux Yacht Enamel is recommended. This is made by International Yacht Paints and is available in yellow (156) and blue (750), which are very close to the recommended BS colours. It can be bought in 250 and 750ml tins from chandlers.

Stencils

Various materials have been used to make stencils of the basic waymark arrow. The Ramblers' Assocation sell a plastic waymarking stencil for 25p, which although not very durable, can be used as a template from which other stencils can be cut. The material must be flexible and easy to clean, but not too thick. Old type vinyl flooring is suitable. Beware of getting paint under the stencil, and spoiling the edge of the arrow. Wipe the back of the stencil between each use.

Some people find it easier to use the stencil to mark in the arrow with a pencil or other pointed object, and then to paint the arrow freehand.

Brushes

The choice depends on personal preference and skill, but a fairly stiff brush of about 6mm is the most popular. A cheap 'fitch' brush as used by schools is suitable.

Aerosols do not give a sufficiently crisp edge to the arrow, although they can be used to give a contrasting background if the arrow does not show up clearly, or if one wishes to obliterate an old waymark, and paint a fresh one on top. Brown should be suitable, although white is useful as a background to blue arrows in woodland.

The sponge 'printers', recommended by the Countryside Commission, do not give a clear enough impression, but can be retouched with a brush to give a sharp arrow.

WAYMARKING KIT

a   Emulsion paint, decanted into a screwtop jar or small tin.

b   Stencil.

c   Brushes, one for each colour. Keep them in small containers of water to stop them drying out between stops.

d   Rag, for wiping brush clean and correcting mistakes.

e   Wire brush and 'Surform' for cleaning surface of object on which arrow is to be painted.

f   Secateurs or other clearing tools, to cut back any overhanging growth which obscures the waymark.

g   Pencil.

It is very important to stand back at least ten paces to judge the angle of the waymark. This is easiest done with two people, one holding the stencil, and the other lining it up from a distance. People working on their own can use a cut-out arrow temporarily attached with a pin or Blutack. It is very easy to get the angle wrong if this check is not made.

The sequence of working will depend on the layout of the paths, the amount of work to be done, and the weather conditions. Each arrow should have two coats of paint. In good drying conditions it may be possible to do both coats on the outward trip, which saves time. As waymarking is necessary from either direction, it should be possible to paint the undercoat of one arrow, then the undercoat of the arrow for the opposite direction. The first arrow will then have dried sufficiently for the topcoat to be applied. In damp weather it will be necessary to work from one end of the path doing the undercoat, and then retrace one's steps to do the topcoat.

OBJECTS

Stiles, gate posts, fences and walls are the most suitable objects on which to place waymarks. Trees can be used if the bark is smooth. The surface should be vertical, smooth, and as permanent as possible. Try and put the waymarks at a uniform height and to the same pattern, for example always on the upper part of a fence post above the top strand of wire.

Signposts

Many fingerposts and stone or concrete signs now incorporate the yellow or blue arrow. This helps get the waymarking system accepted. However, the waymark should only be included if the path itself is waymarked throughout. The Lake District National Park have a useful system: their wooden signs all have a routed (incised) arrow. This is painted in only if the path itself is waymarked.

A second waymark within sight of the signpost helps verify the system, and acts as a substitute if the signpost is vandalised.

Gate posts and stile posts

These are usually the most suitable locations. Put the waymark on the latching gatepost so that it is not obscured if the gate is open. Do not paint the waymark on the gate itself, as its position may move! The upper part of the stile post or the handhold is the obvious position on a stile. It is not usually recommended to paint the waymark on the step of the stile, as the horizontal surface is more quickly weathered or obscured by dirt. It does have the advantage though that awkwardly angled paths can be clearly directed.

Walls

Surfaces need to be thoroughly wire brushed to give good adhesion. The BS yellow is close to the colour of the yellow encrusting lichens, so try and choose a lichen-free part of the wall.

Trees

Beech, ash, sycamore and other smooth-barked trees are suitable if wire brushed and lightly surformed. Birch and other trees with textured or rough bark are not suitable. Cut back overhanging growth so the waymark is clearly visible in all seasons.

Telegraph and electricity supply poles

Permission to use these must be sought from the relevant authorities. Some may give 'blanket permission' for their area, others may refuse or only allow for specific locations. As the poles are heavily creosoted when new, emulsion paint does not adhere well until the poles have aged and dried out a little.

Deterrent notices

It is useful to try and place a waymark within view of a deterrent notice such as 'Private Drive'.

Avoid in particular:

a Unstable objects such as rotten posts, dead branches or loose stones.

b Poorly lit positions such as the trunks of very leafy trees.

c Objects that are low down, and will be obscured by summer growth.

d Surfaces which attract growth of moss or algae, and the natural run-off channels where water runs down tree trunks.

e Gates, which may be left open, thus hiding or altering the direction of the arrow.

Type of Terrain

Farmyards, fields and gardens

Routes through farmyards are often troublesome, both for owner and user. If the owner wishes to make a diversion, this must be referred back to the local authority. It is usually better to have several rather than the minimum of waymarks in a farmyard, as they can be obscured by machinery, stacked fodder and so on.

Great care must be taken that waymarks show clearly which side of a hedge or wall a path takes, or damage to them is likely to result as walkers climb over when they reach a field corner with no gate or stile.

Well kept stiles are the best form of 'waymark' in cultivated land, but arrows are also needed if the network is complex.

Woodland

It is in woodland that waymarking is most valuable, because routes are often winding and not easy to follow, and animal tracks and other paths may make a confusing network through which the right of way is not obvious. Once off the right of way, it is of course difficult to relocate it. Waymarking may initially have to be done as 'blazes', so that each waymark is visible from the previous one. Future maintenance can then select out the important ones, as use becomes established.

In some scrubby woodland or coppice it may be difficult to find branches wide enough for painting a clear arrow. In these cases, a single band of colour painted around the branch is the best solution, and will be understood as long as the waymarking is continuous.

Large fields, downland and heath

The problem here is the lack of suitable objects on which to place waymarks. It is especially difficult on large arable or pasture fields, where waymarking posts (see p176) are not practical because they get in the way of farming operations. Where the field rises away from the walker and no boundary is visible, the route is difficult to follow even with an initial waymark to start the walker in the correct direction, as one soon veers off without a visual target to aim at. The Countryside Commission suggest that large targets, possibly white circles, be used, so that at least once a boundary comes into view, the route can be easily regained. On arable land, the best solution is for paths to be reinstated after ploughing, or not ploughed at all, according to the legal restrictions on the path (see p12). Permanent pasture or parkland can be marked with posts, but these must be very firmly embedded to avoid being disturbed by stock.

Urban areas

It is not easy to make waymarking both sufficiently conspicuous and acceptable in villages and towns. There is usually no problem finding one's way into a village along a public path, but paths out are often difficult to find, and are not easy to distinguish from private paths. Signposting is probably the best method, including signs leading, for example, down side roads to the statutory sign at the junction of the path with a metalled road. This is not practical for linking all paths, but is useful on through routes, and 'unofficial' long-distance or recreational routes (official ones should already be clearly signed).

For the Cotswold Way, permission was gained from Gloucestershire County Council to paint white arrows on the risers of kerbstones, using road-line paint. This avoided having to get separate permissions from numerous landowners, and the system was easily followed by walkers. Disadvantages are that arrows need frequent repainting, and they can be obscured by parked cars.

Mountains and moorlands

There is some controversy over the subject of waymarking paths in the uplands. Waymarking is difficult to do both unobtrusively yet effectively, as there are few suitable objects on which to paint arrows. Cairns, which are 'non-directional waymarks', are the usual method. Waymarking can tempt the ill-equipped walker, who may

follow a line of cairns without map or compass, and then become lost when a mist descends. On the other hand, it can be claimed that cairns are an important and traditional safety feature in helping to keep people to the path on dangerous ground or in difficult weather conditions.

Ideally, all walkers in the uplands should be equipped with a map and compass, and be able to use them to keep to their route even in mist. In the absence of this, it is perhaps necessary that certain paths are continuously cairned if there are particular risks to safety. Tradition also plays a part, and if paths have been marked by cairns for many years without evidence of tempting the ill-equipped, there is no reason to remove them.

The placing of cairns on summits, cols, and at path junctions seem acceptable visually and functionally, as they act as confirmatory signs of a path's destination, not its route. They do though have the disadvantage of acting as targets and attracting erosion and litter. The extent of erosion is seen at some mortared cairns and trig points, now left perched on plugs of uneroded ground.

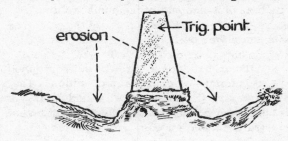

It is suggested that in choosing whether to waymark an upland path, the following principle is applied: If the cairns or posts are to be used to waymark a route across an area where there is no physical or legal reason why people should walk along any one line, it is better not to erect waymarks because:

a   They will tempt the ill-equipped.

b   They are intrusive features in the landscape.

c   If they are followed, a path will be trampled where is may be better that walkers make their own way and cross an area without a path being formed.

If the area is one where it is necessary for physical, legal or safety reasons that walkers should follow one particular line, then waymarking will serve a purpose by helping to create a path and keep it in use. Frequent waymarking posts or arrows are needed at places where restoration work has been done, such as the path to the Devils

Kitchen, at Cwm Idwal in Snowdonia. It is easy to step off the path without realising it, and then have to cross loose scree to regain it, thus causing further damage. In these sorts of places, every turn of the path needs to be waymarked until the path is established and the surrounding scree stabilised.

The summary of this is that if you want a path made and kept to, waymark it; if you want access with no paths, don't waymark.

Waymarking Posts

These are used:

a   Where there are no suitable objects on which to paint waymarks.

b   For marking particular new routes, such as nature trails.

c   As confirmatory markers, particularly on long distance and recreational routes.

Posts can either be:

a   Without directional arrows, but positioned at junctions or changes in direction. The Forestry Commission use round posts with bands of colour to indicate different forest trails in some areas.

b   With directional arrows. The provision of posts has to be carefully worked out in advance to get the sequence of arrows correct, as these are usually routed into the posts and painted in the workshop.

Posts are usually made of 100 x 100mm hardwood such as oak, chestnut or iroko, or preserved softwood. Close-grained wood such as oak or iroko gives a crisper outline to routed letters and symbols. A dark coloured preservative (see p184) is often preferred, as this gives a better contrast to the painted symbols.

The fixing of the post depends on the type of ground and whether vandalism is anticipated. Posts must be fixed securely to resist rubbing by stock. The hole must be bored or dug (see p151) as posts cannot be driven in without damage. Vandalism is discouraged by fixing cross bars to the base of the post, but this requires digging a larger diameter hole in order to fit the post in place. The strongest cross bars are made by knocking metal pipe through pre-drilled holes. Any type of metal pipe is suitable, and can easily

be obtained from dumps and scrap merchants. Alternatively, 150mm nails can be used.

Basic designs are shown below. The height of the post should be chosen to suit the terrain and the vegetation.

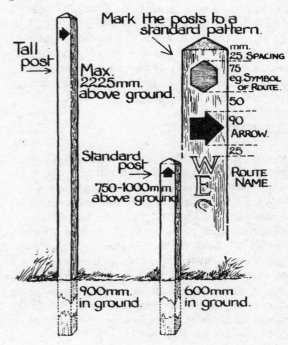

Posts in cairns

In places where it is not possible to dig holes, and a waymark is thought essential, a post can be held in place by a small cairn. The cairn can be dry stone or mortared, depending on the local style and accessibility for bringing materials.

1   Excavate foundation to 150mm if possible. Prise out a hole for the base of the post.

2   Build bottom course, then set post in position, having knocked metal pipes into place. One person holds the post in place.

3   Build up the courses, filling the centre with stones as each course is completed.

4   Finish the top with closely packed stones, pitched on their long axis.

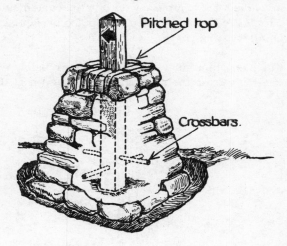

Signposts

Signposts at junctions of paths with metalled roads are the responsibility of the local authority. Some authorities also use them at path junctions. There have been many different types tried over the years, mostly based on the simple 'finger post' principle. It is not intended to discuss the various methods and materials used in making signposts, as this depends on the resources and preferences of the local authority, and volunteers are only occasionally involved in erecting signposts provided by the authority.

The following points are noted:

a   Finger posts of any type are prone to vandalism, as the finger can be broken, bent or twisted out of alignment. However, if correctly positioned they are unambiguous, and are so well known as to be immediately recognised as indicating a public right of way.

b   Metal finger posts, of the so called 'Worboys' type, (after the Worboys Committee on traffic signs who recommended their use) are more vandal resistant than wooden finger posts, but may not be considered as appropriate for use in the countryside.

c   Wooden signposts can be made up by the authority concerned to their own requirements. If desired, wording can be done individually to give destination and distance.

Two of the best designs for wooden finger posts are shown below. They are only vandal-resistant to the extent that their high quality workmanship should engender respect in all but the most malicious vandal.

The Lake District design has finger posts of iroko, a hardwood which gives a neat finish to routed lettering. The fingers are short, so they are difficult to wrench off.

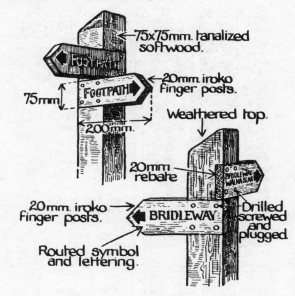

The design below is used by the Countryside Commission for Scotland (see CCS Information Sheet 2.4.16).

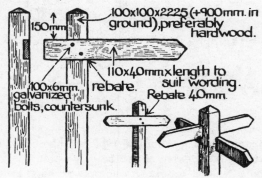

Alternatively, a mortise can be cut in the post, and the finger-post bolted or dowelled in place.

In open country or land used only for grazing, 'low-level' signs can be used. These are no use on ungrazed verges where undergrowth rapidly obscures them, or where they will get in the way of farm machinery. The two examples below are very successful in their particular locations. Slate is also commonly used for signs in Snowdonia and the Lake District, the signs usually being set into stone walls.

The triangular sign is aluminium, and is available either unpainted and unmachined, or painted, machined and complete with plastic-coated steel post. It can also be used to replace broken signs, using existing posts.

The cast aluminium finger sign shown below is available from the same manufacturer (p182 13.1).

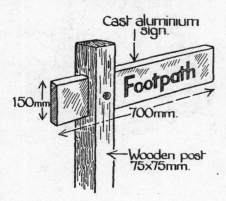

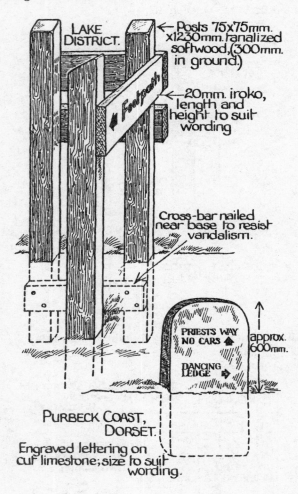

LAKE DISTRICT.

Posts 75x75mm. x1230mm. tanalized softwood, (300mm. in ground.)

20mm. iroko, length and height to suit wording

Cross-bar nailed near base to resist vandalism.

PRIESTS WAY NO CARS ⬆
DANCING LEDGE ➡

approx. 600mm.

PURBECK COAST, DORSET.
Engraved lettering on cut limestone; size to suit wording.

A departure from the usual type of finger post is the triangular design shown below, which is virtually vandal proof. It can only be used to replace a single finger post, and cannot be used at path junctions.

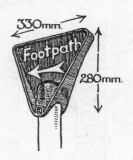

330mm.

Footpath

280mm.

JUNCTIONS

As described above (p172), junctions are difficult to indicate with an easily-understood and vandal-resistant sign. As junctions of three or more paths are quite significant points in the path network, a fairly substantial signpost is appropriate. Some suggestions are given below.

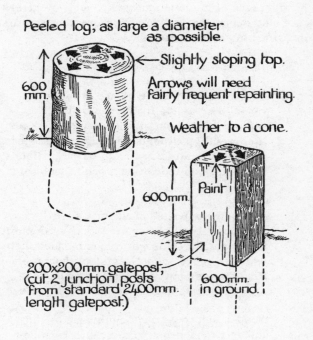

Peeled log; as large a diameter as possible.

600mm.

Slightly sloping top.

Arrows will need fairly frequent repainting.

Weather to a cone.

600mm.

Paint

200x200mm. gatepost; (cut 2 junction posts from standard 2400mm. length gatepost.)

600mm. in ground.

The waymarking post draws attention to the junction, and the direction of the paths is indicated by concrete signs set into the ground. This method is used at Leeds Castle, Kent.

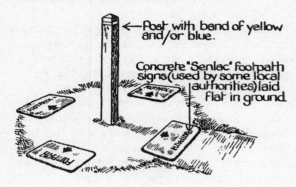

← Post with band of yellow and/or blue.

Concrete "Senlac" footpath signs (used by some local authorities) laid flat in ground.

The information of the top of the cairn must be marked in such a way that it is resistant to vandals. Some suggestions are listed below:

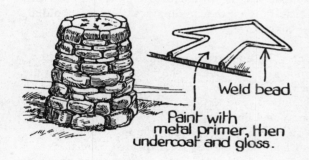

Weld bead.

Paint with metal primer, then undercoat and gloss.

a   Mark arrows with a pointed object into setting in-situ concrete. Use four parts of all-in grit to one part of cement to give a smooth enough texture to make a crisp mark.

b   Set commercially made plastic arrows, destinations and distances into setting concrete. Set sub-flush to reduce vandalism. The Yellow Pages directory will list, under 'Sign makers', many firms able to make arrows and lettering. Alternatively, arrows can be home-made using a resin casting kit (eg Strand Glass).

c   Employ a welder to form arrows and letters of weld bead on 6mm steel plate, or to flame-cut the arrows. Buy the steel at scrap cost if possible. Weld pieces of scrap iron at various angles to the bottom of the plate, and set in top of cairn in 4:1 concrete.

Conservation and the Volunteer Worker

The British Trust for Conservation Volunteers aims to promote the use of volunteers on conservation tasks. In addition to organizing work projects it is able, through its affiliation and group schemes, to offer advice and help with insurance cover, tool purchase and practical technical training.

To ensure the success of any conservation task it is important that the requesting person or agency, the volunteer and the leader all understand their particular responsibilities and roles. All voluntary work should be undertaken in the spirit of the Universal Charter of Volunteer Service, drawn up by the UNESCO Co-ordinating Committee for International Voluntary Service. Three of its most important points are:

1 'The work to be done should be a real need in its proper context and be directly related to a broad framework of development'. In terms of conservation, this means that tasks should be undertaken as integral parts of site management plans, not as isolated exercises. Work should never be undertaken solely for the benefit of the volunteer. Necessary follow-up work after tasks should be planned beforehand to ensure that volunteer effort is not wasted.

2 'The task should be a suitable assignment for a volunteer.' Volunteers cannot successfully tackle all types of work and they should not be used where there is a risk of serious accident or injury, where a financial profit will be made from their labours, where the job is so large that their efforts will have little overall effect, where the skills required are beyond their capabilities so that a bad job results and they become dispirited, or where machines can do the same job more efficiently and for a lower cost.

3 'Voluntary service should not replace paid local labour.' It should complement such work, not supplement it. Employers should make sure in advance that the position of volunteers and paid workers is clear with respect to any relevant labour unions. Further advice may be found in 'Guidelines for the relationships between volunteers and paid non-professional workers', published by the Volunteer Centre, 29 Lower King's Road, Berkhamstead, Hertfordshire HP4 2AB.

Volunteers are rarely 'free labour'. Someone has to pay for transport, materials, tools, insurance, refreshments and any accommodation charges. Before each party makes a commitment to a project it should be clear who is to pay for what. While volunteers may willingly fund their own work, 'user bodies' should be prepared to contribute and should not assume that all volunteers, who are already giving their time and effort, will be able to meet other expenses out of their own pockets. Several grant-aiding bodies may help pay the cost of environmental and conservation projects, notably the Nature Conservancy Council, the World Wildlife Fund and the Countryside Commissions. Details may be found in 'Grant Aid for Voluntary Bodies in the Environmental Field', produced by the Department of the Environment, Room P1/064, 2 Marsham Street, London SW1P 3EB.

It is important that volunteer workers be covered by some sort of public liability insurance for any damage or injury they may cause to property or to the public. Cover up to £250,000 is recommended. Additional insurance to compensate the volunteer for injury to him- or herself or to other volunteers on task should also be considered.

The volunteer group organizer should visit the work site well before the task, to check that the project is suitable and that volunteers will not be exploited, and to plan the best size of working party and the proper tools and equipment. Volunteers should be advised in advance on suitable clothing for the expected conditions. They should be physically fit and come prepared for work and they should genuinely want to volunteer - those 'press-ganged' into service are likely to work poorly, may do more harm than good and may be put off a good cause for life! Young volunteers need more supervision and are best suited to less strenuous jobs, and it is recommended that where they are involved the task should emphasize education. Note that the Agriculture (Avoidance of Accidents to Children) Regulations, 1958, legally restrict the riding on and driving of agricultural machines, vehicles or implements by children under 13 years.

Volunteer group organizers and 'user bodies' both should keep records of the work undertaken: the date of the project, jobs done, techniques used, number of volunteers and details of any notable events including accidents, unusual 'finds', publicity etc. Such information makes it easier to handle problems or queries which may arise after the task. It also provides a background on the task site for future visits, supplies practical data by which the site management plan can be evaluated and allows an assessment to be made of the volunteer effort.

Suppliers

Suppliers are listed chronologically in the order in which they appear in the text. Items mentioned are those for which the firm is particularly recommended, and are not their complete range.

3.1 Leonard Farnell & Co Ltd,
North Mymms, Hatfield, Herts AL9 7SR.
Soil sampling equipment.

3.2 Francis Barker, Fircroft Way,
Edenbridge, Kent TN8 6EX.
Clinometer.

5.1 Bulldog Tools, Clarington Forge,
Wigan, Lancashire.

5.2 Spear and Jackson, St Paul's Road,
Wednesbury, West Midlands WS10 9RA.
All types of tools. Post hole borer.

5.3 T & J Hutton and Co Ltd, Phoenix Works,
Ridgeway, Sheffield.
Eversharp scythette and replaceable blades.

5.4 CeKa, 34 Slough Road, Datchet, Berkshire.
Ratchet secateurs and pruners.

5.5 Wolf Tools, Ross-on-Wye, Herefordshire
HR9 5NE.
Loppers.

5.6 Wilkinson Sword Ltd, Sword House,
Totteridge Road, High Wycombe, Bucks.
Pruners and loppers.

5.7 Sandvik UK Ltd, Manor Way, Halesowen,
West Midlands.
Saws, including folding saws.

5.8 Paice and Sons Ltd, 71/79 Loose Road,
Maidstone, Kent.
Post hole borer.

5.9 C R Bridgedale Ltd, Samuel Street,
Leicester LE1 72B.
Suppliers of Felco pruners.

5.10 Honey Brothers Ltd, New Pond Road,
Peasmarsh, Guildford, Surrey.
Suppliers for tree surgery. Folding saw.
Drill attachment for chain saw.

5.11 Atlas Copco (GB) Ltd, PO Box 79,
Swallowdale Lane, Hemel Hempstead,
Hertfordshire HP2 7HA.
Supplier of Pionjar rock drill.

5.12 Crayford Special Equipment Co Ltd,
Westerham, Kent TN16 1RG.
Argocat.

6.1 Fisher-Humphries Ltd, Wootton Bassett,
Wiltshire SN4 7DB.
Boom-mounted flails.

6.2 Bomford and Evershed Ltd, Salford Priors,
Evesham, Worcs WR11 5SW.
Tractor-mounted flail mowers. Stargrader.

6.3 Twose of Tiverton Ltd, Lowman Works,
Tiverton, Devon.
Tractor-mounted flail mowers.

6.4 Turner International (Engineering) Ltd,
Coughton, Alcester, Warwickshire.
Boom-mounted flails.
Pedestrian flail mowers.

6.5 Link Hampson Ltd, Bone Lane, Newbury,
Berkshire RG14 5TD.
Pedestrian flail mowers.

6.6 Allen Power Equipment Ltd, The Broadway,
Didcot, Oxfordshire OX11 8ES.
Brush cutters.

6.7 G D Mountfield Ltd, Reform Road,
Maidenhead, Berkshire SL6 8DG
Brush cutters.

6.8 Tondu, 33/35 Broughton Street,
Manchester M8 8LZ.
Brush cutters.

6.9 Husqvarna Ltd, PO Box 10,
Oakley Road, Luton LU4 9QW.
Brush cutters.

6.10 G A Edwards Far East Trading Co Ltd,
Bank Street, Melksham, Wiltshire SN12 6LE.
Suppliers of Nikken knapsack mower.

6.11 Murphy Chemical Ltd, Wheathampstead,
St Albans, Hertfordshire.
Tumbleweed, Round-Up.

6.12 ICI Plant Protection, Woolmead House,
Bear Lane, Farnham, Surrey GU9 7UB.
'Garlon' 2.

6.13 May and Baker Ltd, Dagenham, Essex.
Asulox.

6.14 Albright and Wilson Ltd, Industrial
Chemicals Division, PO Box 3, Oldbury,
Warley, Worcs.
Amcide.

6.15 Du Pont (UK) Ltd, Wedgewood Way,
Stevenage, Herts ST1 4QN.
Krenite.

7.1 Brett Plastics Ltd, Speedwell Industrial
Estate, Staveley, Derbyshire.
Plastic pipes.

7.2 Wavin Plastics Ltd, PO Box 12, Hayes,
Middlesex UB3 1EY.
Plastic pipes.

7.3 Armco Ltd, Stephenson Street, Newport,
Gwent NPT 0XH.
Steel pipes.

7.4 Rocla (GB) Ltd, Milton Keynes, Bucks.
Concrete Pipes.

7.5 Hepworth Iron Co Ltd, Hazlehead,
Stocksbridge, Sheffield S30 5HG.
Clay pipes.

7.6 Charcon Products Ltd, Hulland Ward,
Derby DE6 3ET.
Safeticurb.

8.1 ICI Fibres, 'Terram', Pontypool, Gwent.

8.2 Malcolm, Ogilvie & Co Ltd, Constable
Works, Dundee DD3 6NL.
Wyretex.

8.3 Woodland Mulch Ltd, Warren Camp,
Crowborough, Sussex TN6 1UB.

9.1 Mono Concrete Ltd, Epic House, Lower
Hill Street, Leicester LE1 3SH.

9.2 River and Sea Gabions (London) Ltd,
2 Swallow Place, London WIR 8SQ.

9.3 Brooklyns Westbrick Ltd, Concrete Products,
17 Paul Street, Taunton, Somerset TA1 3PF.

11.1 John Chambers, 15 Westleigh Road,
Barton, Seagrave, Kettering, Northants
NN15 5AJ.   Seeds.

11.2 Emorsgate Seeds, Emorsgate,
Terrington St Clement, King's Lynn,
Norfolk PE34 4NY.

11.3 Suffolk Herbs, Sawyers Farm, Lt. Cornard,
Sudbury, Suffolk.

11.4 Johnsons Seeds, W W Johnson and Son Ltd,
Boston, Lincolnshire.

11.5 Dunlop Irrigation Services Ltd, PO Box 1,
Thame Park Road, Thame, Oxon OX9 3RQ.
Dunebond (soil stabiliser).

11.6 Huls UK Ltd, Byrom House, Quay Street,
Manchester M3 3HQ.
Huls 801 (soil stabiliser).

11.7 Vinyl Products Ltd, Mill Lane,
Carshalton, Surrey SM5 2JU.
Vinamul 3277 (soil stabiliser).

11.8 Bridon Fibres and Plastics Ltd, Team
Works, Dunston, Gateshead, Tyne and Wear.
Broplene Land Mesh.

12.1 British Gates and Timber Ltd,
Biddenden, near Ashford, Kent.

12.2 A G Lane Ltd, Roadside Sawmills,
Longhope, Glos GL17 0LP.
Gates and fencing.

13.1 Gascoignes Non-Ferrous Foundries Ltd,
706/707 Stirling Road, Trading Estate,
Slough, Berkshire SL1 4TD.
Aluminium signs.

Timbers and Preservatives

Timber is used for many types of footpath work, and it is important that a suitable timber and method of preservative treatment (if necessary) are chosen.

Timber Characteristics

The following table lists some of the different types of timber widely available in Britain. The properties of the timber of any one species may vary according to the soil and climate in which it was grown.

Working qualities

This refers to the ease of drilling, sawing etc.

Durability

This is the measure of resistance to decay, and refers to untreated timber.

VD   Very Durable
D   Durable
MD   Moderately Durable
ND   Non Durable

Heartwood is generally more durable than sapwood. Footbridges should be made of timber of durability MD, D or VD.

Density

This varies with moisture content. Values given are at 12% moisture content.

Permeability

This measures the extent to which preservative can penetrate a timber. Sapwood is usually more easily penetrated than heartwood. Round timbers are of sapwood on the outside, and with the exception of larch, are easily penetrated.

ER   Extremely Resistant (absorbs only small amount)
R   Resistant (difficult to penetrate more than 3 - 6mm)
MR   Moderately Resistant (6 - 18mm penetration in 2 - 3 hours)
P   Permeable (penetrates without difficulty)

Permeability also depends on the moisture content of the timber. Green (freshly cut) timber contains a higher moisture content and is less permeable than seasoned timber.

Use

These are suggested uses in footpath work.

Species	Working qualities	Durability	Density kg/m³	Permeability	Use
SOFTWOODS					
Douglas fir	Good	MD	529	R	Bridge beams, stiles, gates, posts.
Larch	Good	MD	592	VR (sapwood also VR)	Bridge beams, stiles, gates, posts.
Scots pine	Good	MD	689	R (sapwood P)	Bridge beams, stiles, gates, posts.
Western red cedar	Good	MD	368	R	Gates
HARDWOODS					
Ekki	Difficult	VD	1025	ER	Bridge beams and decking.
Elm	Medium	ND	625	MR	Boardwalks
Iroko	Medium/ difficult	VD	641	ER	Signs, especially for routed letters.
Keruing	Difficult	MD	721	R	Gates, bridge beams and decking.
Oak	Medium	D	625	ER	Gates, stiles, posts.
Sweet chestnut	Good	D	545	VR	Gates, stiles, posts, boardwalks.

For properties of other timbers, see 'Timbers - their properties and uses' (TRADA 1979), from which this table was compiled.

Structural grade

Timber is graded according to the proportion of defects in a sawn section. Softwood grades are:

GS   General Structural Grade

SS   Special Structural Grade

If graded by machine instead of by hand, equivalent grades are termed:

MGS   Machine Structural Grade

MSS   Machine Special Structural Grade

The timber is further defined by species name, or where the species is not significant, by a species group S1, S2 or S3. Species within these groups are as follows:

Species group	Standard name	Origin
S1	Douglas fir	Imported
	Pitch pine	Imported
S2	Western hemlock	Imported
	Parana pine	Imported
	Redwood	Imported
	Whitewood	Imported
	Canadian spruce	Imported
	Douglas fir	Home grown
	Larch	Home grown
	Scots pine	Home grown
S3	European spruce	Home grown
	Sitka spruce	Home grown
	Western red cedar	Imported

S1 or S2 must be used for footbridges, and are recommended for all other uses where strength is important.

Suitable hardwood grades for footbridges are M50 or M75.

Preservative Treatments

There are three factors to be considered:

1 The permeability of the timber (see above)
2 The method of application
3 The type of preservative

METHODS OF APPLICATION

Application of preservative is better done on timber which has already been cut to size.

Brushing and spraying

This is the least effective method, and should only be used on joints and end grain cut after main treatment when no other method is possible. Apply liberally, using at least four coats.

If this is the only possible method of application, timber should be re-treated every few years to give satisfactory penetration. Creosote penetrates better if heated before application.

Immersion

Usually called 'dipping' for periods up to about 10 minutes, and 'steeping' if immersed for hours or days. This is only effective for permeable timber, as resistant timber absorbs little even after long periods of steeping. Organic solvents or creosote can be used for dipping. Creosote is normally used for steeping and is a convenient method for the treatment of estate timber which is often permeable sapwood, such as round fence posts.

It is impossible to give optimum immersion periods but as a general rule aim at a minimum of five minutes for dipping, and 24 hours for steeping. Dipping is the best method where timber, for example newly cut end grain, has to be treated on site.

Hot and cold open tank

Timber is immersed in cold preservative (usually creosote), and then heated to 85-95° C and kept at this temperature for one to three hours. The timber is left in the tank while it cools, during which period most of the absorption takes place. This is more effective than simple immersion, and is a useful do-it-yourself method for treating estate timber. For further details of this method see MAFF Fixed Equipment of the Farm leaflet 17.

Double vacuum

This is a commercial process for applying organic solvents. It gives long term protection against decay and insect attack for timber not in contact with the ground.

Pressure

This is a commercial process, and the most effective method of applying preservative. Even resistant timbers will absorb sufficient preservative. All wood merchants should be able to supply pressure-treated wood, and may also treat wood supplied and cut by the customer.

TYPES OF PRESERVATIVE

Types of preservative	Possible application methods in order of effectiveness	Properties
Coal tar creosote to BS 144	1 Pressure 2 Open tank 3 Brush/spray	Helps prevent checking (cracking) of timber exposed to the weather. Highly toxic to wood-destroying fungi and insects. Non-corrosive to metals. Gives some repellance to water which helps retard dimensional movement. Strong smell, making treated timber unpleasant to handle. Stains. Cannot be painted over. Inflammability increased for short period after treatment, but once thoroughly dried, is slightly more fire-resistant than untreated timber.
Water-borne salts, mostly copper/chrome/arsenic to BS 4072 eg Tanalith C Celcure A Wolman CCA	1 Pressure	No smell. Forms insoluble compound in treated wood which cannot be leached out. Gives greenish-grey colour to timber. Can be painted over and glued. Not moisture repelling. Timber swells during treatment and should be re-dried to restore to original size before use as joinery. Non-corrosive to metals, non-staining and non-flammable.
Organic solvent, An organic fungicide/ insecticide in an organic solvent. eg Cuprinol Solignum	1 Double vacuum 2 Immersion 3 Brush/spray	Preservative resistant to leaching but some lost by evaporation. Non-corrosive and non-staining. Can be painted and glued. Treatment does not cause swelling of the timber and therefore can be used on accurately cut wood. No fire hazard once solvent has evaporated. Not water-repellent, but additives can be included to resist moisture changes in use. Many colours available. Not suitable for timber in contact with the ground. Suitable for signboards, handrails, gates.

Untreated Timber

The heartwood of D or VD timber is not usually treated as it is naturally resistant to decay.

Occasionally timber may be used for footpath work immediately after felling; for example as bridge beams (see p106), or for building revetments or steps on a woodland path where timber is to hand. Treatment with preservative is not possible because of the high moisture content of the timber. The saving on the cost of supply must be balanced against its reduced resistance to decay, compared with treated timber.

Concrete

HEALTH AND HYGIENE

Fresh cement is caustic and rapidly damages unprotected skin. It may also cause dermatitis. Use a gauntlet style of PVC glove, and avoid gloves with cotton cuffs.

CARE OF TOOLS AND MIXER

Concrete sets rapidly and extra water hastens the process. Do not leave tools to 'soak' in water. Keep a bucket of water and a brush handy, and wash tools completely once an hour and before every break.

After emptying a mixer, put the water for the next batch in straight away, and leave the mixer running. A shovel of aggregate will help keep the drum clean. On hot days wash the mixer out before lunch. At the end of the day start to clean out the mixer well before work finishes. Use plenty of water and aggregate to clean the drum. The blades may need knocking occasionally with a hammer to remove encrustations. Wash off the mixer with water and a brush. Check the oil and fuel. Leave in the emptying position and remove the cleaning aggregate as it will otherwise be set by morning!

MIXING

Strength of mix

Most concrete for small jobs is made of ¾" 'all-in' aggregate and Portland cement. If the all-in aggregate (often called gravel) is too 'sharp' (ie not enough fine material) then add some sand.

Mixes are given aggregate : cement

Mix	Properties	Use
1:1	Very strong and quick setting; may crack when curing.	Small repairs in existing work
2:1		
3:1	Strong concrete; may surface crack when curing.	Structures needing strength, or those that will be exposed to hard frosts.
4:1		
5:1	Good concrete, but takes a while to reach fair hardness.	Leanest mix (ie lowest proportion of cement) suitable for structures exposed to the weather.
6:1		
7:1		Suitable for concreting in posts etc if no movement likely for three to four days.
8:1		
9:1 to 12:1	A weak mix, but will reach fair strength after a couple of weeks.	For backfilling of pipes across tracks where either buried deep, or no traffic likely for a couple of weeks.

Use tap or spring water, but not stream or sea water which contain impurities that will interfere with setting. Use as little water as possible while achieving a workable mix. Only a fraction of the water in the mix is used in the chemical reaction that hardens the concrete. The rest has to escape by evaporation, leaving tiny voids that weaken the concrete. Because the moisture content of the aggregate varies, the exact proportion cannot be specified. If making several batches for one job, use a consistent proportion of water or the strength of the concrete will vary.

The smaller the proportion of cement in the mix, the drier it can be made. For 4:1 - 6:1, a well mixed batch should flow when tipped from a mixer, and form a hump about 150mm high in a filled wheelbarrow. Don't overfill the barrow, as movement will flatten the hump!

Buried concrete takes water from the ground and needs little or no water in the mix if full strength is not needed immediately. For example posts and pipes can be concreted in using a dry mix.

Machine mixing

Do not overfill; half a wheelbarrow of aggregate is sufficient.

1. Start the engine and load the aggregate.
2. Add the cement and allow to mix thoroughly.
3. Add some of the water and mix for about two minutes.
4. Check the consistency; add extra water sparingly until the mix is workable.
5. Continue mixing for another two minutes, then unload into wheelbarrow.

Hand mixing

1. Measure the aggregate onto a board, and add the measured cement.

2. Mix thoroughly by turning the heap over and forming a new heap. Repeat three times.

3. Flatten to form a pile about 150mm high, and make a large hole in the centre. Add some water.

4. Gently turn the edges of the pile into the centre. Turn the whole pile, sprinkling on more water if necessary.

STORAGE

Cement has a short life. Buy what you need just before you use it. Keep bags of cement off the ground and concrete floors by resting them on pallets or planks. Keep rain off the bags.

Wheelbarrows make good 'roofs' for single bags on site.

SHUTTERING AND FORMWORK

Shuttering and formwork must be strong, water-tight and simple to dismantle. There are cheap 'shuttering grades' of plywood available, and less commonly, shuttering nails with a special head for easy removal.

Paint the wood with shuttering oil to prevent the concrete sticking.

POURING, VIBRATING AND FINISHING

Pour the concrete gently to avoid putting stress on the shuttering. Use a vibrating poker to remove the air from the mix. If you don't have a poker, use a stick and work hard! Then hammer the sides of the shuttering with a club hammer to further shake down the concrete.

If there are poor patches or gaps when the shuttering is removed, render them up with 4:1 sharp sand to Portland cement.

Listed below are various finishes suitable for small jobs:

Tamped

Use a board across the setting concrete to level and consolidate the top (see p83). Use either as final finish, which gives good grip for the feet, or as preparation for any of the following methods.

Steel float

After tamping, a steel float can be used to give a very smooth finish. Requires skill to use effectively.

Wood float

After tamping, use a wood float to give a good roughish finish. This looks effective and is much easier to achieve than a finish with a steel float, especially on dryish mixes.

Brushed

When the concrete is just 'going off', brush with a hard broom to leave stones exposed. This finish can also be made using a hosepipe.

CURING AND PROTECTING

In the summer, cover work with damp hessian, and keep damp for as long as possible. Rain should be kept off wet concrete, but once 'gone off', rain is good for it. In the winter, protect from frost by covering with boards, hessian, straw or any other available material.

Concrete takes about 48 hours to be strong enough for shuttering to be removed, two to three weeks to gain reasonable strength, and about 100 years to reach maximum hardness.

Conservation and Amenity Organisations

General

Association for the Protection of Rural Scotland
    20 Falkland Avenue, Newton Mearns,
    Strathclyde G77 5DR

Biological Records Centre
    Monks Wood Experimental Station,
    Abbots Ripton, Cambridgeshire PE17 2LS

Botanical Society of the British Isles
    c/o Department of Botany, British Museum,
    (Natural History), Cromwell Rd, London SW7

British Ecological Society
    62 London Road, Reading, Berkshire

British Herpetological Society
    c/o The Zoological Society of London,
    Regents Park, London NW1 4RY

British Trust for Conservation Volunteers
    Headquarters: 36 St Mary's Street,
    Wallingford, Oxfordshire OX10 0EU

    London: 2 Selous Street, Camden Town,
    London NW1

    South: Hatchlands, East Clandon, Guildford,
    Surrey GU4 7RT

    Thames and Chilterns: 36 St Mary's Street,
    Wallingford, Oxfordshire OX10 0EU

    South West: Newton Park Estate Yard,
    Newton St Loe, Corston, Avon

    East Anglia: Bayfordbury House, Hertford,
    Hertfordshire

    Midlands: 577 Bristol Road, Selly Oak,
    Birmingham 29

    South Yorkshire: Balby Road,
    Doncaster, South Yorkshire

    West Yorkshire: Hollybush Farm, Broad
    Lane, Kirkstall, Leeds, West Yorkshire

    North West: 40 Cannon Street, Preston,
    Lancashire

    North East: 423 Chillingham Road,
    Heaton, Newcastle upon Tyne 6

    Scotland: 70 Main Street, Doune,
    Perthshire

    Wales: Forest Farm, Forest Farm Road,
    Whitchurch, Cardiff

British Trust for Ornithology
    Beech Grove, Tring, Hertfordshire

Cement and Concrete Association
    Wexham Springs, Wexham, Slough SL3 6PL

Civic Trust
    17 Carlton House Terrace, London SW1

Convention of Scottish Local Authorities
    3 Forres Street, Edinburgh EH3 6BL

Council for National Parks
    4 Hobart Place, London SW1W 0HY

Council for the Protection of Rural England
    4 Hobart Place, London SW1W 0HY

Council for the Protection of Rural Wales
    14 Broad Street, Welshpool, Powys SY1X 8PQ

Country Landowners Association
    16 Belgrave Square, London SW1X 8PQ

Countryside Commission (England and Wales)
    John Dower House, Crescent Place,
    Cheltenham, Gloucestershire GL50 3RA

Countryside Commission for Scotland
    Battleby, Redgorton, Perth PH1 3EW

Dartington Amenity Research Trust
    Shinners Bridge, Dartington, Totnes, Devon

Farming and Wildlife Advisory Group
    The Lodge, Sandy, Bedfordshire SG19 2DL

Field Studies Council
    62 Wilson Street, London EC2A 2BU

Forestry Commission
    231 Corstorphine Road, Edinburgh EH12 7AT

Friends of the Earth
    377 City Road, London EC1V 1NA

Game Conservancy
    Burgate Manor, Fordingbridge, Hampshire

Geological Society of London
    Burlington House, Piccadilly, London W1V 0JU

Institute of Terrestrial Ecology
    68 Hills Road, Cambridge CB2 1LA

Landscape Institute
    12 Carlton House Terrace,
    London SW1Y 5AH

Access to the Countryside

Association of Countryside Rangers
c/o Mike Marshall, 2 Causeway Cottages,
Middleton, Saxmundham, Suffolk

British Horse Society
National Equestrian Centre, Kenilworth,
Warwickshire CV8 2LR

British Mountaineering Council
Precinct Centre, Booth Street East,
Manchester

Byways and Bridleways Trust
9 Queen Anne's Gate, London SW1H 9BY

The Chiltern Society
Membership Secretary, 9 Kings Road,
Berkhamsted, Herts

Commons, Open Spaces and Footpaths
Preservation Society (The Open Spaces Society)
25a Bell Street, Henley on Thames, Oxon

Ramblers' Association
1/5 Wandsworth Road, London SW8 2LJ

Scottish Rights of Way Society
28 Rutland Square, Edinburgh EH1 2BW

Trail Riders' Fellowship
Rights of Way Officer, 39 Warren Road,
Thorne, Doncaster, South Yorkshire

Youth Hostels Association
Trevelyan House, St Albans, Herts

National Association of Local Councils
100 Great Russell Street, London WC1B 3LD

National Council for Voluntary Organisations
26 Bedford Square, London WC1

National Farmers Union
Agricultural House, Knightsbridge,
London SW1

National Farmers Union of Scotland
17 Grosvenor Crescent, Edinburgh EH12 5EH

National Federation of Young Farmers' Clubs
National Agricultural Centre, Kenilworth,
Warwickshire CV8 2LG

National Trust
42 Queen Anne's Gate, London SW1H 9AS

National Trust for Scotland
5 Charlotte Square, Edinburgh EH2 4DU

Natural Environmental Research Council
Polaris House, North Star Avenue,
Swindon, Wiltshire SN2 1EU

Nature Conservancy Council
19 Belgrave Square, London SW1X 8PY

Ordnance Survey
Romsey Road, Maybush, Southampton SO9 4DH

Royal Society for Nature Conservation
The Green, Nettleham, Lincoln LN2 2NR

Royal Society for the Protection of Birds
The Lodge, Sandy, Bedfordshire SG19 2DL

Scottish Landowners' Federation
18 Abercromby Place, Edinburgh 3

Scottish Wildlife Trust
25 Johnston Terrace, Edinburgh EH1 2NH

Sports Council
70 Brompton Road, London SW3 1EX

Timber Research and Development Association
Hughenden Valley, High Wycombe, Bucks

Tree Council
35 Belgrave Square, London SW1X 8QN

Woodland Trust
Westgate, Grantham, Lincs.

World Wildlife Fund
Panda House, 11-13 Ockford Road,
Godalming, Surrey

Bibliography

This list includes works referred to in the text, and others relevant to the subject of footpaths.

Ansell, Judith (1979) — Public Path Waymarking Project Report 1976-79, The Ramblers' Association

Appalachian Trail Conference (1980) — Appalachian Trail Fieldbook, Appalachian Trail Conference

Baily, J C (1981) — Parish Councils and Footpaths, Cumbria Association of Local Councils

Bayfield, N G (1973) — Use and deterioration of some Scottish hill paths, Journal of Applied Ecology 10, pp 635-644

Bayfield, N G (1980) — Replacement of vegetation on disturbed ground near ski lifts in the Cairngorm Mountains, Scotland, Journal of Biogeography 7, pp 249-260

Birchard, William and Proudman, Robert D (1981) — Trail Design, Construction and Maintenance, Appalachian Trail Conference

Bishop, O N (1973) — Natural Communities, John Murray

British Standards Institution (1979) — Specification for stiles, bridle gates and kissing gates BS5709:1979 amended 1982

Bradshaw, A D and Handley, J — Low cost grassing of sites awaiting redevelopment, ILA Techniques No 11, Landscape Institute

Brooks, A (1977) — Dry Stone Walling, British Trust for Conservation Volunteers

Brooks, A (1980) — Woodlands, British Trust for Conservation Volunteers

Brotherton, I, Maurice, O, Barrow, G and Fishwick, A (1977) — Tarn Hows – an approach to the management of a popular beauty spot, Countryside Commission CCP 106

Brown, Andrew C H (1974) — The Construction and Design of Signs in the Countryside, Countryside Commission for Scotland

Bryant, J D (1978) — Pony Trekking Project Report – Working Paper, Brecon Beacons National Park

Burch, William R Jr (edit) (1979) — Long Distance Trails, School of Forestry and Environmental Studies, Yale University

Campbell, Ian (1974) — A Practical Guide to the Law of Footpaths and Bridleways, Commons, Open Spaces and Footpaths Preservation Society

Campbell, Ian and Clayden, Paul (1980) — The Law of Commons and Village Greens, Open Spaces and Footpaths Preservation Society

Clapham, A R, Tutin, T G and Warburg, E F (1968) — Excursion Flora of the British Isles, Cambridge University Press

Clayden, P and Trevelyan J (1983) — Rights of Way: a guide to law and practice, Ramblers' Association and Open Spaces Society

Colville, Ian (1978) — Review of Selected Footpath Construction and Maintenance Work in Upland Scotland. Published by the BTCV for the Countryside Commission for Scotland

Convention of Scottish Local Authorities and the Countryside Commission for Scotland (1978) — Walks and Footpaths for recreation – A course of action, Countryside Commission for Scotland

Countryside Commission — Self-closing gates, Management and Design Note 5

Countryside Commission (1971) — Pennine Way Survey CCP 63

Countryside Commission (1973) — Surfacing Materials for use on Footpaths, Cycle-tracks and Bridleways CCP 66

Countryside Commission (1974) — Upland Management Experiment CCP 82

Countryside Commission (1974) — Waymarking for Footpath and Bridleway